W9-CEI-347

FLAVOR

FLAVOR

The Science of Our Most Neglected Sense

BOB HOLMES

W. W. NORTON & COMPANY
Independent Publishers Since 1923
New York | London

Printed in the United States of America
First Edition

For information about permission to reproduce selections from this book,
write to Permissions, W. W. Norton & Company, Inc.,
500 Fifth Avenue, New York, NY 10110

For information about special discounts for bulk purchases, please contact
W. W. Norton Special Sales at specialsales@wwnorton.com or 800-233-4830

Manufacturing by LSC Communications Harrisonburg
Book design by Chris Welch
Production manager: Louise Mattarelliano

Library of Congress Cataloging-in-Publication Data

Names: Holmes, Bob (Evolutionary biologist), author.
Title: Flavor : the science of our most neglected sense / Bob Holmes.
Description: First edition. | New York : W.W. Norton & Company, [2017] |
"Independent Publishers Since 1923." | Includes bibliographical references and index.
Identifiers: LCCN 2016046897 | ISBN 9780393244427 (hardcover)
Subjects: LCSH: Taste. | Taste buds. | Flavor. | Senses and sensation.
Classification: LCC QP456 .H66 2017 | DDC 612.8/7—dc23
LC record available at https://lccn.loc.gov/2016046897

W. W. Norton & Company, Inc.
500 Fifth Avenue, New York, N.Y. 10110
www.wwnorton.com

W. W. Norton & Company Ltd.
15 Carlisle Street, London W1D 3BS

1 2 3 4 5 6 7 8 9 0

For Deb, my partner

in flavor and in life

CONTENTS

FLAVOR

INTRODUCTION

Have you ever wondered why beer and salted peanuts go so well together? Scientists know the answer: Salty tastes inhibit bitter ones, so the nuts tame the beer's bite and allow some of its other flavors to step forward. Once you know this principle, you can apply it in many other ways. Serve the nuts (or pretzels) with gin and tonic. Add a little extra salt if tonight's broccoli is especially bitter. Put a pinch of salt on your morning grapefruit.

The science of flavor is full of insights like that, but hardly anyone knows about them. That's because flavor barely registers in the screenplay of our daily lives. We rarely examine the flavors we experience, and as a result we don't know how to talk about them or think about them. Here's a thought experiment to prove it: Take a moment and bring to mind one of your favorite pieces of music. Recall how it's put together and what makes it special for you. Is it the subtle use of the saxophone in the bridge section? The way the first violin and cello trade the theme back and forth? The moment of breath-holding suspense just before the vocals start? Chances are, you can put your finger on several specific elements that make that music sing for you. You can name the instruments that are playing, you can pick out the melody, bass line, and vocals, you know how fast the beat is.

Now try to describe your favorite apple variety in the same detail. Why do you like, say, Fujis better than any other? Most likely, you'll stammer out a few generalities about crispness or sweetness or "more flavor." But unless you're a trained apple taster (and such people do exist), you probably won't be able to manage much more than that. You certainly won't be able to name the apple's flavor elements as nimbly as you named the instruments in your favorite music, and you probably won't have much to say about how the flavor profile of each bite builds and ebbs.

And our imprecision is not confined to just apples. Can you describe how the flavor of halibut differs from red snapper? Or how Brie cheese differs from Cheddar? The fact is that for most of us, flavor remains a vague, undeveloped concept. We say "dinner tasted good," or "I like those peaches," but we never dip beneath the surface of those superficial responses. It's not that we're blind to flavor. If you can recognize that a Fuji apple differs from a Spartan, or that Brie differs from Cheddar—and almost all of us can—you have the basic perceptual tools to explore the world of flavor in greater depth.

What holds most of us back is that although we experience flavor every day, we just don't know much about it. We sip our morning coffee or enjoy our dinner while largely ignorant of the complex interplay of taste, smell, touch, sight, and even expectation that creates the sensation we know as flavor. Without that knowledge, we lack the means to describe what we experience, and as a result, far too often we simply don't notice the fine details of what we eat and drink. It's as though the entire world of flavor has been relegated to the background—elevator music for the palate, as it were.

Sometimes that's fine, of course. Sometimes all we really want is background music, or a quick bite to eat without bothering too much about the details. But in our musical world, most of us take that extra step now and then. We pay attention and dig a little deeper, and our lives are much richer for it. We can have the same rich experience in our flavor lives, too—but only if we learn more about the world of flavor: how we perceive flavor, where it comes from, and how we can maximize it, both on the farm and in our kitchens. That's where this book can help.

Paying attention to flavor makes life not just richer but deeper, because flavor appreciation may be a uniquely human gift. The biology of our species—the fact that we live in social groups, inhabit essentially every environment on Earth, and eat a diverse, omnivorous diet—means that our ancestors had to become very good at certain skills. They had to recognize faces to tell friend from foe, neighbor from relative, and honest dealer from cheater. As a result, all of us, with a few rare, pathological exceptions, are indeed skilled at picking out the subtle differences that distinguish one face from the next. We recognize, and often remember, the face of someone we went to school with years ago, and the stranger we met casually at a party yesterday. And we do it instantaneously, at a glance, not by laboriously piecing together evidence from nose, ears, cheekbones, and eyes. This recognition skill is special and unique to faces. It's not just a consequence of sharp perception and attention to detail—we have nowhere near the same ability to recognize people by their hands, for example.

Flavor recognition is another of humans' special skills. As omnivores, our ancestors had to judge what they could eat and what they couldn't, and flavor is how they made that decision. Those skills are now part of our evolutionary heritage. "All humans are flavor experts in the same sense that we're face experts," says Paul Breslin, a leading psychologist who studies flavor perception. "It is literally a life-or-death matter. If you eat the wrong things, you're dead." We recognize the flavor of a strawberry or a pineapple or a green bean in a flash, even if we can't put a name to it without prompting.

In fact, our flavor sense may have played a large role in making humans into the species we are. Anthropologist Richard Wrangham argues that we could never have evolved our huge, expensive brains without the easy calories made available by cooking. Raw foods simply don't yield enough calories to get our modern, big-brained bodies through the day. Our cousins the chimps spend hours each day laboriously chewing their raw foods to extract the calories—time and energy that humans can put to better use. And people who follow a raw-food diet typically lose significant weight, even with blenders and juicers to take the place of constant chewing. Cooking breaks down indigestible tissues into smaller, more digestible fragments, and thus helps us get more from our meals for less effort. And in the process, it creates a whole host of delicious new flavors.

We are also the only species that seasons its food, deliberately altering it with the highly flavored plant parts we call herbs and spices. It's quite possible that our taste for spices has an evolutionary root, too. Many spices have antibacterial properties—in fact, common seasonings such as garlic, onion, and oregano inhibit the growth of almost every bacterium tested. And the cultures that make the heaviest use of spices—think of the garlic and black pepper of Thai food, the ginger and coriander of India, the chili peppers of Mexico—come from warmer climates, where bacterial spoilage is a bigger issue. In contrast, the most lightly spiced cuisines—those of Scandinavia and northern Europe—hail from cooler climates. Once again, our uniquely human attention to flavor, in this case the flavor of spices, turns out to have arisen as a matter of life and death.

Our unusual anatomy cooperates in making humans connoisseurs of flavor. Our upright posture and oddly shaped head (compared with other mammals) helps our noses focus less on smells coming from the outside world and more on the flavors wafting up from the food in our mouths. And flavor engages a disproportionate share of our big, powerful brains. When you enjoy a delicious piece of cheese, or a glass of wine, or a cookie, you're engaging more brain systems than for any other behavior. Flavor taps into sensory systems for taste, smell, texture, sound, and sight. It involves motor systems for coordinating the muscles that allow you to chew and swallow. It activates the unconscious linkages that regulate appetite, hunger, and satiety. And, not least, it fires up the higher-level thought processes that help you identify, evaluate, remember, and react to what you're eating. That's a big bundle of brain activity from a simple bite of food.

Flavor pulls on our brains in subtle but powerful ways. When odor information—the most important component of flavor—enters the brain, it goes directly to the ancient parts of the brain responsible for emotion and memory. It doesn't reach the conscious, logical part of the cerebral cortex until several stops later. That's the neuroscientific basis for flavor's remarkable ability to move us: A taste of a favorite food can transport us back to our childhood more powerfully than a song or a photo ever could. It's no accident that Marcel Proust's seven-volume *Remembrance of Things Past* was sparked by the flavor of a madeleine, or tea cake. That emotional pull may also explain why immigrants hold on to the flavors of their native country long after they've adopted new languages, new modes of dress—even, sometimes, new religions. Food binds ethnic groups together across generations and across oceans and national boundaries. We so often use flavors as ethnic markers, with the treasures of one culture being seen (at least initially) as disgusting by others. The French have their stinky cheeses, the Americans their peanut butter, the Australians Vegemite, and the Japanese the mucilaginous fermented soybeans known as *natto*.

For many of us, venturing outside our own ethnic markers is one of the best bridges into another culture. "I've been to many countries in the world, and one of the things I've done in every country is visit food markets," says Breslin. "I've never really thought about why that is, but I can't imagine not doing it. It's always been a rewarding experience." Most people share that response to some degree. Who, after all, would take a trip to Italy and eat only at McDonald's, or live on pizza in China?

The roots of flavor, it seems, run deep into the human condition. But flavor also spices our daily life. All of us have to eat every day, and most of us seek out tastier foods when we have the choice. Grocery shoppers consistently report that flavor is their main guide in deciding what to buy each week, trumping considerations of health, price, and environmental impact. And people rate the pleasure of a fine meal higher than sports, hobbies, reading, or entertainment. Only holidays, sex, and family time ranked higher. And when asked why that fine meal is so pleasurable, more people cite flavor than any other reason.

For millions of people, the act of cooking a daily meal is a creative, rewarding experience. If you've picked this book off a bookshelf, you're probably one of that group. I know I am. We read cookbooks, trawl the Internet for interesting new recipes, and gradually build our kitchen repertoires. Yet most home cooks approach flavor haphazardly. We do what the recipe says, or what we've always done. Sometimes we mix things up by following our intuition and tossing in a handful of basil or sprinkling a grating of nutmeg. But we're just following instructions, or intuition, or tradition; we lack the deeper understanding of flavor that could give shape to our efforts. In a way, we're like the self-taught guitarist who can copy riffs by ear but can't read music and has no formal training in harmony. We bumble along pretty well, and occasionally stumble on something that works beautifully. But think how much more we could accomplish with a better understanding of what we're doing.

For an eye-opening (palate-opening?) demonstration of how little most people know about flavor, take what I call the jelly bean test. Get hold of some jelly beans or other candies that come in a mix of flavors. The fancy, many-flavored jelly beans you can buy everywhere these days are ideal, but a tube of rainbow-flavored Life Savers would work just fine, too, or Jolly Rancher hard candies. It doesn't matter—the important thing is just that you have several flavors to choose from. Now close your eyes, pinch your nose, and have a friend hand you one of the candies. Pop it in your mouth—still pinching your nose—and pay attention to the flavor. Not much there, right? You'll get the sweetness of the sugar, of course, and maybe a little sourness or saltiness, depending on the candy. But what flavor is the jelly bean? You won't be able to tell.

Now release your nose, and see how the flavor suddenly explodes into your mouth. What was once merely sweet and a bit sour is now suddenly LEMON! or CHERRY! What's changed is that you've brought your sense of smell into the game. The lesson here is that what appears to be a simple taste perception is more complex than we realize: Even though we refer to the "taste" of the jelly bean, taste itself is the least important part of the equation. Most of the flavor we actually experience is the result of smell, not taste. (For an even more vivid illustration of this point, hold your nose and try to tell the difference between a cube of apple and a cube of onion. It's harder than you'd think.)

The English language contributes to the confusion. We have separate nouns, "taste" and "flavor," but we use them in greatly overlapping ways. Decades ago, psychologist Paul Rozin found that English speakers generally use taste when they're referring to sweet, sour, salty, and bitter, which—along with the less widely known umami—form the five basic tastes that our tongue can detect. But we use taste and flavor almost interchangeably to refer to the bigger picture—the whole jelly bean, if you will. And when it comes to verbs, we make no distinction at all, using taste for everything, all the time. We say that dinner tasted good and mean much more than merely that it was properly salted and not too bitter. Indeed, when we have a cold we say we can't taste anything—even though, in fact, taste is all we have left when our nose is plugged. One word, two meanings—that just about guarantees confusion. We also have the verb "savor," but it doesn't help much. To savor something usually implies that we ate with pleasure. No one would say, "I savored dinner and didn't like it." (Other languages are no better. Rozin polled native speakers of nine other languages and found that most use just a single word to cover both taste [in the strict sense] and flavor. Only two—French and Hungarian—have two different words, and even the French blur the distinction somewhat.) There's no easy solution to the confusion. Throughout *Flavor*, I do my best to be clear about whether I'm talking about a taste or a flavor, but I fall back on the verb "taste" for both. I hope the context clarifies which meaning I intend.

In fact, flavor has even more dimensions than just taste and smell. Every one of our five senses—taste, smell, touch, sound, and even sight—contributes meaningfully to the way we perceive flavor. The best way to think about flavor is that it is the sum of all the sensations we get when we have food in the mouth. That leads to some surprising discoveries: the weight of a bowl, the color of a plate, the crunch of a potato chip, and even the choice of background music can have a direct effect on how we perceive flavor.

The meals we cook and the foods we eat are more than just a daily source of pleasure, of course. They also affect our health in profound ways. That's especially true now, when poor diet and excess calories have fed an epidemic of obesity that threatens, for the first time in centuries, to shorten our life expectancy. More Americans are overweight than not, and the rest of the Western world is catching up quickly. Many experts point to our taste for sweetened soft drinks and high-fat, high-carbohydrate, high-calorie fast food as a primary cause.

Once again, that puts flavor at the center of the picture. If we're to do something about obesity, as individuals and as a society, we'll need to understand why we eat what we eat. We'll need to know how flavor drives our food choices, and whether we can use it as a lever to shift our consumption patterns. We'll need to understand how flavor helps tell us when we're full, and whether we overeat when meals are especially tasty. These turn out to be complex questions that scientists don't fully understand yet, but some of the answers they've found may surprise you.

Until recently, a book exploring the science of flavor would have been much shorter and more limited in scope. Within the past few years, however, scientists have made huge strides toward understanding every step of the pathway from food to perception to behavior. It's no exaggeration to say that the science of flavor is one of the fastest-moving and most exciting disciplines around these days. A large proportion of the hundreds of scientific papers I read in the course of research for this book are just a year or two old. No doubt even more big discoveries await in the next few years. And as a bonus, it's science that everyone can relate to, because it's about the foods we eat every day, the pleasure we take in a glass of wine, a mug of beer, or a cup of coffee, and the question every one of us faces every day: What would I like to eat for dinner?

In the early 1990s, biologists Linda Buck and Richard Axel identified the receptors responsible for detecting odor molecules, work that earned them a Nobel Prize in 2004. With receptors finally in hand, and aided by the human genome sequence completed early this century, other researchers are racing to crack the code by which the nose encodes the many different smells—possibly many millions—that compose the flavors of the foods we eat. Others are identifying the chemical receptors that detect a chili pepper's fire and the cool of mint. Even the five basic tastes that we've known for a century are having to share the tongue with at least one, and possibly several, other tastes, as we'll see.

As scientists refine our understanding, we're coming to realize that every person on the planet lives in their own unique flavor world defined by their genetic endowment, their upbringing and later food experiences, and the culture in which they live. We're beginning to learn how these unique flavor worlds help define some of our strong likes and dislikes for certain foods. Take for example, former U.S. president George H. W. Bush's famous distaste for broccoli. ("I do not like broccoli," Bush told reporters back in 1990. "And I haven't liked it since I was a little kid and my mother made me eat it. And I'm president of the United States, and I'm not going to eat any more broccoli!") We can't know for sure without testing the former president's genes, but it's a pretty good bet that Bush carries a particular genetic variant of one specific bitter taste receptor, which makes broccoli and other mustard-family vegetables taste especially bitter to him. Your own genes undoubtedly shape your food preferences in similar ways—although genetics is not destiny: not everyone who tastes the bitterness hates it.

From our senses to the kitchen, flavor is a much deeper and more complex subject than most people realize. In these pages, I provide what you can think of as a user's guide to your flavor senses. By the end, I hope you'll have a better understanding of what flavor is, how we perceive it, and how we can use that knowledge to enjoy a richer flavor experience.

Flavor is a book for anyone who enjoys flavor—that is, for almost anyone. You don't have to be a flavor virtuoso to find a deeper appreciation of what's on your plate or what's in your glass. I'm certainly no virtuoso. I'm just an amateur cook of middling ability and above-average enthusiasm, with a nose of roughly average ability. If I can find my way into a world of high-definition flavor, anyone can.

Chapter 1

BROCCOLI AND TONIC

As a journalist, and a classically polite Canadian, I don't often stick my tongue out at the people I'm interviewing. It seems bad form, somehow. But I'm doing it now to Linda Bartoshuk, the grande dame of taste research. Fortunately, she doesn't seem to mind.

"Oh, your tongue is gorgeous," she gushes. She leans close and paints the tip of my tongue with a Q-Tip dipped in blue food coloring, which highlights the taste buds on my tongue. (To be accurate, they're not actually taste buds, which are microscopic. Those mushroom-shaped bumps on the surface of the tongue that most people call taste buds are, technically speaking, really fungiform papillae, an impressive-sounding Latin term that means "mushroom-shaped bumps.")

I hold up a mirror to see what Bartoshuk sees on my tongue. Tiny pink islands stand out in a sea of blue dye. "You see those red dots on the front? Those are fungiform papillae," she says. "You have a lot. Oh, and you have them all the way back! You're very close to a supertaster."

Understanding this notion of the supertaster—that some people have a much more acute sense of taste than others—is what has brought me to Bartoshuk's lab here at the University of Florida in Gainesville. It was Bartoshuk who first suggested, back in 1991, that people tend to fall into three groups, based on their ability to taste a bitter compound known as propylthiouracil, or PROP.

You may have encountered PROP in a high school biology lab or at a science museum somewhere. You're handed a little piece of filter paper infused with a modest amount of PROP, which you put on your tongue. Some people—the nontasters—just shrug, tasting basically nothing apart from filter paper. Others—the tasters—notice an unpleasant bitter taste, while the third group experiences extreme bitterness. This third group, the supertasters, are easy to recognize: They're the ones who make an anguished face and rush off to find something—anything—to wash that horrible taste out of their mouth. Bartoshuk often asks people to rate the intensity of PROP's bitterness on a scale from 0 to 100, where 100 is the most intense sensation they've ever experienced—the pain of childbirth, say, or a broken bone, or the visual sensation of looking directly at the sun. Supertasters often rate the bitterness of PROP in the 60–80 range, nearly in broken-bone territory. Sure enough, I'd score it a 60: nasty, but not debilitating. "That's into supertaster territory," says Bartoshuk. "That's in the area where you're not screaming, but you're definitely much higher than normal, and your tongue looks it."

And it's not just bitterness. Supertasters tend to rate sweets as sweeter, salt as saltier, and chili peppers as hotter. They even report that food aromas are more intense, says Bartoshuk—probably because taste and smell reinforce each other in the brain.

Before I get too smug about my taste acuity, though, Bartoshuk points out that supertasters tend to be pretty boring eaters. Most of them prefer to avoid the intense taste experiences that come with highly flavored foods, so their diets are often bland and narrow. (I knew a man once who lived on a habitual diet of lima beans and milk. I would bet good money he was a supertaster.) In particular, bitter greens and other vegetables don't show up very often on the plates of most supertasters.

That's where I start to get confused, because that doesn't sound like me. I love collard greens, rapini, and other bitter vegetables; I always pick the hoppiest beer I can find; I drink my coffee black and without sugar; tonic water is my soft drink of choice—indeed, the only one I ever drink. In contrast, Bartoshuk—a nontaster—has very pronounced food aversions. She detests tonic water, for example. "When I first tasted it, I couldn't believe it was a beverage," she says. "I cannot stand greens. The bitter taste is just beyond belief to me."

So what's going on? It's time to look more closely at this whole supertaster notion, which turns out to be more complex than it appears at first glance.

A little background: Even though we talk loosely about "tasting" complex foods like wine and cheese, most of their flavor actually comes from our sense of smell. In fact, even though we usually treat smell and taste as one and the same, they actually have different jobs to do. Smell is all about identification—it answers the question, "What is it?" It tells you the difference between rosemary and oregano, Brie and Stilton, or Cabernet Sauvignon and Pinot Noir. It tells you when something is burning on the stove, and it tells you that the dog needs a bath. We can even recognize the odor of our own bodies and those of our sweethearts.

Taste, in contrast, answers a different question: "Do I want to eat this?" Taste is all about broad categories of good and bad, the yes/no, red-light/green-light decisions that would have been so crucial for our hunter-gatherer ancestors. As omnivores without access to grocery stores, they had to make these calls every day, and our taste repertoire bears witness. Everyone knows the "four basic tastes": sweet, salty, sour, and bitter. If you've been paying attention the past few years, you've probably heard of a fifth: umami, a Japanese term that means "delicious flavor" and is usually translated as "savory," "brothy," or "meaty." (There might be additional basic tastes, too, as we'll see.) A closer look at each of those five tastes reveals a lot about what was important to our ancestors.

Sweet tastes, most obviously, mark the presence of sugars, an important source of calories. Even starchy foods such as potatoes and grains yield a hint of sweetness as we chew, because enzymes in our saliva break down the starches into sweet-tasting sugars. Umami comes from amino acids—in particular, one called glutamate, though others contribute as well—that indicate the presence of proteins, another major class of nutrients. And our taste for salt would have helped our ancestors identify the electrolytes that were so precious and hard to find before salt shakers sat on every table. Hardly surprising, then, that we're hardwired, even as infants, to be drawn to sweet, umami, and salty tastes.

But taste also warns us when we're about to eat something that might be harmful. Many toxins taste bitter, so we're hardwired to reject bitter foods. Just watch the face of a toddler who unknowingly sips from a glass of tonic water—or, for that matter, an adult who gets surprised by a bitter-tasting berry or a first taste of aquavit or Fernet-Branca. The bitterness triggers our poison-avoidance reflex, and we make a "yucky face," sticking out the tongue in a reflex that pushes the threatening food out of the mouth. Similarly, we tend to reject sourness, which could signal spoilage or unripe, indigestible fruit. With experience, and practice, we often learn to override that hardwiring for certain foods—coffee, hoppy beers, brussels sprouts, sour candy—but few, if any, people like them right away. Remember your first sip of coffee?

Other species, with narrower diets, have fewer decisions to make and can often get by with fewer tastes. In the use-it-or-lose-it world of evolution, that often means they lose those extraneous tastes. Cats, for example, are entirely carnivorous, so they would never need to recognize high-sugar foods—and, in fact, they seem indifferent to sweetness. Sure enough, when researchers looked more closely, they found that cats have lost a crucial gene that would allow them to taste sweetness. Other carnivores, such as otters, sea lions, and hyenas, have also lost the ability to taste sweet. In each case, a different genetic defect was responsible, suggesting that the taste for sweet has been lost several different times on the evolutionary tree—presumably, each time an omnivorous ancestor switched to an exclusively carnivorous diet. In contrast, pandas, which eat nothing but bamboo, have no need to detect protein in their diet and have lost the taste for umami. Other scientists recently discovered an even more extreme example of taste loss: vampire bats, with their blood-only diet, live in a taste world focused so tightly on recognizing the saltiness of blood that they lack the ability to taste sweet, umami, or bitter.

And by the way, while we're talking tastes: You've no doubt seen one of those "taste maps" of the tongue that purports to show that we taste sweet at the tip, salty and sour along the sides, and bitter at the back. If you're up to date on your reading, you may also have heard that it's completely wrong. As it turns out, though, both sides of the debate are guilty of a little exaggeration. There do seem to be minor differences in sensitivity to the various tastes across the tongue, with some regions a little more sensitive to sweet and others a little more sensitive to bitter, but the differences probably don't matter much. And you can easily verify that the tastes aren't tightly segregated into distinct zones, simply by dipping a Q-Tip in salt water and painting the tip of your tongue. You'll taste the saltiness, even though you're in what's supposed to be the "sweet" zone. Best to just forget the whole notion of the taste map.

Those five basic tastes seem like a pretty unimpressive set compared with the vast array of aromas we encounter in our food. Is taste really all that important to us, or is it only a minor part of our flavor experience? To answer that question, I headed from Bartoshuk's lab in Florida to the Monell Chemical Senses Center in Philadelphia.

You could think of Monell as the Vatican City of flavor research, but without the fancy architecture. The nondescript brick office building, on the fringes of the University of Pennsylvania campus just west of downtown, could house anything: doctors' offices, accountants, engineers. Only a giant bronze sculpture of a nose and mouth, on a concrete plinth next to the front door, hints that something more unusual is inside: one of the planet's greatest concentrations of researchers on the basic biology of the flavor senses.

Inside, Monell's boardroom looks much as you'd expect for such an august institution: long, dark wood table polished to a high gloss, high-backed leather chairs, off-white walls hung with framed memorabilia and interesting-but-not-too-interesting art. It all adds up to a clear message: significant discussions of important ideas take place here.

Over the years, many of those ideas have come from Gary Beauchamp, the center's longtime director. (He stepped down in 2014.) Beauchamp is a small, dapper man with silver hair, a neatly trimmed goatee, and a dignified manner. It's easy to imagine him charming a sizable check out of a deep-pocketed donor. Right now, though, he's leaning back in his chair at the head of the table, gazing thoughtfully at the ceiling. "Glaarglglglgl," he says gently.

"Glaarglglglgl," we all gargle in response. We each lean forward to spit into a plastic cup, then wipe stray droplets from lips and face.

This peculiar boardroom meeting had its genesis at a conference three months earlier, where I met Beauchamp for the first time. We'd been talking about the relative importance of taste versus smell in determining flavor. Most experts come down rather heavily on the side of smell as carrying the lion's share of flavor, since it carries so much more information than just sweet, sour, salty, bitter, and umami. Some say smell accounts for 70 percent of flavor; others put it at 90 percent or more.

But Beauchamp wasn't buying it. In fact, he disagreed vehemently when I suggested this at the conference. "Clearly, olfaction is very, very important," he said. "But the idea that it's 70 percent of flavor is complete bullshit, in my view." Olfaction gets all the attention, he went on, because we all know what it's like to lose the sense of smell. Anyone who's ever had a head cold knows that a plugged nose makes food bland and tasteless (though in fact, "tasteless" is actually the exact opposite of the truth—what you're experiencing is taste alone, in isolation, with smell taken out of the equation). And the jelly bean test gives an even more dramatic demonstration, because it's so quickly reversible.

On the other hand, most of us have never had the inverse experience, since nothing in our everyday life can take away the sense of taste while leaving smell intact. There is no reverse jelly bean test where you can hold your tongue to keep yourself from tasting. Doctors, too, often see patients who have lost their sense of smell as a result of head injury, viral infection, or just as a consequence of aging. By contrast, relatively few people lose their sense of taste. The big exception is cancer patients who undergo radiation to their head and neck, which often damages taste receptors and nerves. And their experience tells a terrible story, said Beauchamp, whose wife's uncle was one of the unlucky ones: as bad as it is to lose your sense of smell, losing taste is far, far worse. "When people lose their sense of taste, they don't eat. They starve themselves to death," he said. "My view is that taste is absolutely the bedrock of flavor."

Moreover, Beauchamp thought he had a way to test that claim—an experiment that would be, in effect, something fairly close to a reverse jelly bean test. Certain drugs, it turns out, can block the perception of salt and sweetness, two of the most important tastes in many meals. "When those things are gone, my guess would be that your dinner would be absolutely awful," Beauchamp said. He'd taken the salt-blocking drug before, out of curiosity, but had never tried knocking out both tastes at once. We agreed that it would be an interesting test to try sometime.

Which brings us back to the Monell boardroom, several months later. Beauchamp, two of his colleagues, and I are gargling with chlorhexidine, an over-the-counter mouthwash sometimes used to treat gum disease, which has the odd side effect of blocking the taste of salt. Each of us tosses back four little cough-syrup cups of the bitter-tasting stuff, one after another, swishing each around in our mouth for thirty seconds and gargling occasionally to make sure the solution reaches well back in the throat, before spitting it out. We follow that with four more cough-syrup cups of a swampy-flavored tea made from a South American plant called *Gymnema*, which knocks out sweet taste.

Sure enough, all that gargling and swishing seems to have obliterated those two tastes. A sip of Pepsi yields a brief prickling on my tongue—the mouthfeel, or touch, sensation from the carbonation—then its flavor vanishes completely. I dip my finger in salt crystals and lick it off: nothing, except a tiny residual saltiness at the very back of my throat where the chlorhexidine didn't quite reach. Now we turn to our experimental "lunch," a burger and fries from the food truck parked in front of the building, now quartered into individual servings. Without the most important parts of our sense of taste, would we be able to stomach the meal, or would we, like Beauchamp's wife's uncle, just give up?

Sure enough, downing the burger is like eating a mouthful of textured clay or soft plastic pellets. Have you ever accidentally left the salt out of homemade bread and been bored by the blahness of the resulting loaf? This burger is like that, but much more so—and we've only knocked out two of the five basic tastes. When I eliminate smell, too, by pinching my nose shut, it's even worse: a totally nondescript experience. But even the loss of taste alone is really crippling—much worse than doing without smell, as I have when eating a burger while nursing a cold. So it looks like the burger, at least, bears out Beauchamp's theory that taste trumps smell.

The fries, though, aren't so bad—partly because I get a little residual saltiness at the back of my tongue, where the gargle didn't reach, but partly because there's still something interesting going on when I put them in my mouth. Could this be the "fat taste" that many researchers now think belongs in the canon, or the fat's pleasant mouthfeel? Then, too, the ketchup still gives a pleasantly tart/umami kick, though it's oddly altered by the lack of sweetness.

All in all, I think Beauchamp might be right. If I had to pick one flavor sense to lose, I'd probably rather give up smell and keep taste. Food that lacks the basic tastes is not actively bad or repugnant, just utterly unfoodlike. If every meal was like this, I'd certainly have a hard time sitting down to eat three times a day.

You'd think that such a vital sensory system—especially one that's relatively simple, with only a handful of basic tastes—would be completely understood by now. Not so: Huge gaps remain in our understanding of how taste works. Scientists can't even agree on how many basic tastes there are.

At the simplest level, we know a fair bit about some parts of the story. Tasting happens when the thing we taste—the tastant—binds to receptors on taste cells on the tongue or palate. The tastants for salty and sour—sodium and acids, respectively—go right into the taste cells and activate them, in a process that's still not fully understood. The process is pretty well worked out for sweet, umami, and bitter, though, so let's look at them in a little more detail.

Leo Tolstoy famously wrote that happy families are all alike, while each unhappy family is unhappy in its own way. Taste is kind of like that, too. The good tastes, umami and sweet, are each recognized by a single receptor, a two-part protein that's woven into the outer membrane of taste cells. (There may be other, unrelated receptor molecules that are also sensitive to sweet and umami tastes, but the evidence there isn't conclusive yet.) For umami, the two parts are called *T1R1* and *T1R3*, while for sweet it's *T1R2* and *T1R3*. The amino acid glutamate, or one of several sugars, slip into fitted pockets on these combo receptors. The traditional metaphor here is a key fitting into a lock, but you could also think of the way an expensive camera slots into a foam carrying case. If you have the wrong case for the camera, it doesn't fit. If they match, the camera slips in perfectly.

The bad, bitter taste, on the other hand, makes use of a huge committee of receptors called *T2R* receptors. Each member of the committee—there are at least twenty-five in humans—handles a different range of bitter compounds. Some, like *T2R10*, *T2R14*, and *T2R46*, are what the scientists like to call "promiscuous," mating with a wide range of bitter compounds. In fact, if you just had those three *T2R*s and no others, you'd be able to detect more than half of a test sample of 104 diverse bitter-tasting chemicals. Some other bitter receptors, like *T2R3*, seem to be monogamous, with only a single chemical known to activate them. It works the other way, too: Some chemicals activate many different *T2R*s, while other chemicals trigger just a single bitter receptor. What's more, bitter receptors seem to come and go over the course of evolution: The human genome is littered with the rusting hulks of bitter receptor genes that no longer function. These must have been important in our evolutionary past, but—like the sweet receptors of cats—they have become irrelevant enough that we no longer need them, and we haven't noticed their absence.

Scientists still don't know whether all those bitter receptors send identical signals to the brain—in which case there's just a single taste we call "bitter"—or whether we can actually taste the difference among different classes of bitter. Part of the problem is that when you compare, say, the bitter of a hoppy ale to that of coffee, you're not just comparing the output of particular bitter receptors, particularly *T2R1* for hops and *T2R7* for caffeine. Instead, you're really comparing the whole flavor profile of the two drinks. Even if you hold your nose while drinking—which few of us do in a social setting—the two differ in other tastes such as sweet and sour. We don't generally experience, let alone compare, pure bitter tastes in our everyday lives. But research scientists do, and at least one expert is convinced that there's more than one bitter taste. "When you do a lot of bitter research, and taste these things side by side, you realize they taste different," says John Hayes, a flavor researcher at Penn State University. And that plays out in our food preferences, he thinks. "I like my beers very hoppy," he continues. "I love a good IPA. And yet I can't stand grapefruit, because I find it bitter. If there was only one kind of bitterness, then the learning process that I went through to learn to like my IPAs presumably would have generalized over to grapefruit juice. And the fact that it hasn't generalized, to me, starts to provide some of the argument for why there's more than one kind of bitterness." He's now hard at work in his lab trying to prove this hunch.

There's a lot left to learn about umami, too. Now that scientists have found the receptor responsible for umami, there's little doubt that it deserves to be considered the fifth basic taste. But for most people, it's still a little hard to accept. After all, everyone knows exactly what you mean when you talk about sweet, salty, sour, or bitter. If umami is just as fundamental a taste, then why does it so often need to be explained? What makes umami so obscure?

Two reasons, says taste researcher Paul Breslin of Monell. First, we routinely experience the other tastes in nearly pure form: the sweetness of honey, the sourness of lemon juice, the bitterness of radicchio, a pinch of salt. "You get, like, a pure shot of those," he says. "But you're never going to experience pure glutamate in the world. You're not going to find a pile of it that you can lick. We really only experience it in combination with a lot of other things."

That inability to isolate the taste of umami is reinforced by Breslin's second reason: Our umami receptors max out at low intensity, so we're physically unable to experience *very* umami in the same way we can taste *very* salty or *very* bitter simply by piling on the salt or brewing a cup of extra-strong espresso. Thanks to our perceptual apparatus, umami can never be anything more than a subtle sensation. It's as though you could understand the color red from referring to a crimson rose, yellow from a lemon, and green from a midsummer forest, but then had to try to figure out blue from skim milk.

There's also a cultural component to our umami blindness, however. Most people from Western countries struggle to put a name to umami taste sensations, but that's not the case for people from Asian countries. "If you look at Japanese kids, you put MSG in their mouth and they say 'umami' like that," says Danielle Reed, Breslin's colleague at Monell, with a snap of her fingers, "like American kids put sugar in their mouth and say 'sweet.'" As umami becomes a more prominent part of our food culture—with food writers tossing the term around freely now, and restaurants like Umami Burger part of the general conversation—it's likely that our umami blindness will gradually recede into the past.

When that happens, it will be interesting to see whether our appreciation of umami rescues the reputation of MSG. MSG—

monosodium glutamate—is, after all, merely sodium, which is pure salty taste, and glutamate, which is pure umami taste. When chefs work hard to enhance umami by adding dashi or soy sauce to stocks, incorporating mushrooms in stews, aging their meat, or incorporating fermented ingredients, they are simply boosting the glutamate content of the finished dish—and we love the result. Why, then, do so many of us shudder at the thought of boosting glutamate directly, by adding it in pure form? We routinely see signs in restaurant windows or labels on packaged foods proclaiming "No MSG!" But what self-respecting cook would ever boast of doing without salt, sugar, or lemon juice?

The reason for MSG's bad reputation, of course, is that many people feel that it causes an unpleasant reaction when they eat food that has had MSG added. This notion, which is now commonplace, is actually a relatively new idea. It first appeared in 1968 when a Chinese American doctor, Robert Ho Man Kwok, published a letter in a leading medical journal describing "numbness at the back of the neck, gradually radiating to both arms and the back, general weakness and palpitation" beginning a few minutes after starting a meal at a Chinese restaurant. Kwok wasn't sure what caused this "Chinese restaurant syndrome," but he suggested MSG as one possibility.

The news media quickly picked up the story, and similar anecdotes began popping up all over. Soon researchers began giving MSG to volunteers, who reported symptoms similar to Kwok's syndrome and added others, such as headache, to the list. The idea that MSG could be bad for you became widespread. Soon Ralph Nader and others were urging governments to regulate its use.

But even then, skeptics wondered: If MSG really produces such unpleasant symptoms, why didn't anyone notice this sooner?

After all, the food industry had been using MSG for decades, and not just in Chinese food. By the time Kwok published his letter, the United States alone produced fifty-eight million pounds of MSG every year, and it showed up in everything from baby food to canned soup to TV dinners. Yet no one had remarked on a "TV dinner syndrome" or "canned soup syndrome."

All this made MSG research a hot topic during the 1970s. As scientists dug deeper into the compound's effects, though, Chinese restaurant syndrome began to look more and more iffy. The most damning evidence came from several studies of people who claimed to be sensitive to MSG. Researchers gave all the volunteers a capsule to swallow, without telling them whether it was MSG or a dummy capsule containing inert ingredients. (Using a swallowable capsule prevented the volunteers from tasting the difference.) If there was any truth to their self-professed sensitivity, participants should have developed symptoms of Chinese restaurant syndrome when they consumed MSG, but not when they took the placebo. In fact, though, the volunteers reported just as many symptoms with the placebo as with MSG—strong evidence that their symptoms stemmed from what they expected to happen, rather than from what they actually ate.

That's not as surprising as it sounds. Most of us have felt a little funny after eating now and then. Maybe you ate a bit too much, or too fast, or were feeling tense for other reasons. And many of us are especially cautious after eating something new, as Chinese food would have been for many people in the 1960s. Once one unsettling experience has planted a seed of doubt, our expectations can start to turn our future responses into a self-fulfilling prophecy.

In fact, when researchers looked back at the early studies that first suggested a link between MSG and Chinese restaurant syndrome, most of them suffered from this expectation problem. Usually, the researchers had not bothered to hide the taste of MSG, so that participants in the study could probably guess whether they'd consumed MSG or a placebo. Some studies didn't even attempt a placebo but simply gave MSG to people and asked them if they felt any symptoms—an ideal situation for expectations to take the driver's seat.

Even so, there are no doubt a few people out there with a genuine sensitivity to MSG. But if pure MSG causes problems, those people should also have trouble with dishes containing mushrooms, soy sauce, Parmesan cheese, and other foods naturally rich in umami flavor. And of course, overuse of MSG could bring its own problems, just like overuse of salt or lemon juice or any other seasoning. With those cautions in mind, though, there's no reason why most cooks shouldn't incorporate MSG into their repertoire of seasonings. After all, most kitchens have recourse to pure chemical seasonings to boost the tastes of salt (sodium chloride), sweet (sucrose), and sour (acetic acid, aka vinegar). Why not keep a little MSG on hand for those times when a dish needs a boost of pure umami?

When it comes to industrial taste research, though, umami is small potatoes. The big money is in sweet. Like umami, sweetness is all about a single taste receptor, as far as we know (although, as we'll see in a moment, there may be some reason to suspect other receptors, too). And that simplicity has sparked a huge effort from scientists—mostly working for Big Food—to find alternative ways to tickle that receptor that aren't accompanied by the caloric charge of real sugar.

Most of the artificial sweeteners already on the market are the result of pure dumb luck. The oldest was discovered by accident in 1878 when Constantin Fahlberg, a chemist working on coal tar products in Baltimore, forgot to wash his hands before supper and noticed that his bread tasted "unspeakably sweet." He thought nothing of it until he noticed the same sweetness on his napkin, his water glass, and, eventually, his thumb. Fascinated, Fahlberg dashed back to the lab and started tasting everything he could find. Fortunately, he found the sweet compound, which we now know as saccharin, before he got to anything too toxic.

Cyclamate has much the same story: in 1937, a chemist at the University of Illinois set his cigarette down on the corner of his lab bench and noticed when he picked it up again that it tasted sweet. Aspartame: a chemist working on antiulcer drugs in 1965 licked his finger to help pick up a piece of paper and noticed a sweet taste. Sucralose: a chemist in London was asked by his boss in 1976 to "test" a new chemical, but misheard it as "taste"—a potentially lethal error for a chemist, but one that worked out well for the company.

Artificial sweeteners reduce calories for two reasons. Some, such as saccharin and sucralose, are not broken down by the body and thus provide no calories. Others, such as aspartame, taste sweet at lower concentrations than regular sugar, so even though they are digestible, they deliver their sweetness with fewer calories. There's a catch—even though some of these chemicals start tasting sweet at low concentrations, their sweetness often maxes out early, too. No matter how much saccharin you dump in your coffee, for example, it never tastes sweeter than a 10.1 percent sugar solution. That's a problem for soft drink manufacturers, because regular Coke is 10.4 percent sugar, and Pepsi is about 11 percent.

That's not the only reason artificially sweetened drinks taste a little weird to many people. Another is that most of the artificial sweeteners trigger not just the sweet receptor but also one of our many bitter receptors, producing a bitter aftertaste that many people find highly objectionable. Since people have different sets of bitter receptors, some of us are bothered by certain sweeteners and not others. I get a bitter taste from saccharin, for example, which suggests that my *T2R31* bitter receptor works well. On the other hand, I get no bitter taste from the low-calorie natural sweetener stevia, so I probably have a broken version of whichever bitter receptor (still unknown) responds to that sweetener.

But bitterness isn't the only problem with the taste of artificial sweeteners. Linda Bartoshuk, for example, can't taste the bitterness of aspartame or saccharin, yet she knows them when she tastes them. "The sweet of saccharin is nothing like the sweet of sucrose. I don't know how anybody could ever confuse them," she says. "And if I accidentally get a beverage with aspartame in it, I'm not confused for a moment. I don't like it. So it's pretty clear that not all sweets are the same."

Part of the reason for that is that each sweetener has its own distinctive timing for triggering the sweet receptor. Real sugar reaches its peak sweetness in about four seconds, then the taste trails off about ten seconds later. Most artificial sweeteners hang on too long, producing a cloying aftertaste. Aspartame, for example, starts a second later and lasts four seconds longer. But Bartoshuk thinks the taste differences might also point to the existence of a second kind of sweet receptor, as yet unknown. It's hard to believe that we don't know everything yet about something as obvious—and as lucrative for Big Food companies—as sweetness, but there you are.

If artificial sweeteners are the king of taste research, dollar-wise, then salt substitutes would have to be the queen. The average American consumes about 9 grams of salt daily, almost half again as much as the recommended maximum of 5.8 grams per day, and the majority of that comes from processed foods. That high salt intake is a big reason why sixty-five million American adults have high blood pressure. As a result, food-processing companies are under a lot of pressure to find ways of reducing the sodium in their products.

The problem is, that's not easy to do. As anyone who's spent time in a kitchen knows, salt contributes much more than just a salty taste to the flavor of a dish. Used judiciously, salt can enhance all the other flavors, making meat meatier, beans beanier, and potatoes potatoier. That's largely because the sodium ions help draw other flavor compounds—mostly components that enhance smell, not taste—out of the ingredients and into solution, where we can detect them. Omit the salt, and your food literally has less flavor. This explains why a skilled cook can often tell by smell whether a dish needs more salt.

To find out how food scientists are working around the problem, I asked Peter de Kok, who works for the food science company NIZO in the Netherlands. De Kok—who, like most Dutch scientists, speaks flawless English—comes across as a cheerful fellow with a boundless enthusiasm for salt reduction. There are three ways to deliver all the flavor bang of regular salt with less sodium, he says. You already know about the first one if you've ever bought "low-sodium salt" in the grocery store: simply replace some or all of the sodium with another salt ion. The more chemically similar your replacement is to sodium, the better job it does of substituting. In practice, that pretty much restricts the choice to potassium, which is about 60 percent as salty as sodium. (Lithium would actually be a better substitute, flavor-wise, but it has powerful psychological effects—just ask anyone with bipolar disorder.) Unfortunately, many people—though I'm not one of them—also get a bitter taste from potassium, so companies can only swap out part of the sodium in their low-sodium salt.

If you don't want to replace sodium with a different ion, a second approach is to find a way to get more flavor from the same amount of salt. Smaller salt crystals dissolve more quickly, so they taste saltier when sprinkled atop food. (The converse is also true, of course—when you eat a pretzel topped with the traditional big salt grains, you're actually getting more sodium than necessary for the amount of salty flavor it delivers.) De Kok and his colleagues also try to find ways of getting more of a food's sodium out of the food and into your mouth, where you can taste it. For example, they've been working on changing the texture of sausages to make them juicier. In essence, he says, when you chew these juicier sausages you squeeze more of the salty moisture out into your mouth, so the sausages taste just as salty with 15 percent less salt. Yet another strategy exploits the value of contrast: They've patented a method of making bread with alternating layers of salted and unsalted dough. As you bite through the layers, the contrast makes the salty parts stand out, so that the whole bread tastes about 30 percent saltier than it otherwise would.

The third way to cut back on salt without reducing flavor is a bit more devious: Trick the brain into thinking the food is saltier than it is. As we'll see, your brain blends aromas and tastes together into a unified perception of flavor. Knowing this, de Kok and his team have been experimenting with adding aromas that we're used to smelling in high-salt contexts. Because anchovies are typically salty, for example, you mentally "add salt" whenever you get a whiff of anchovies, whether the salt is really there or not. You can't flavor everything with anchovies, though, so de Kok found an alternative that's more universally beloved but still salty: bacon. The researchers isolated about two dozen different aroma compounds from bacon, then tested each one to see if it enhanced people's perception of saltiness. Sure enough, they found three that did. By selecting meat that's naturally high in those three compounds, de Kok's team was able to make sausages that still tasted right but used 25 percent less salt.

Of all our flavor senses, taste is the one most tightly identified with the mouth. Yet even that is a little misleading: Now that scientists know what several of our taste receptors look like, they're finding them all over the body—in our guts, in our brains, even in our lungs. Taste, it seems, plays a wider role than we thought, though many of the details still aren't clear.

The best known of these "other" taste receptors are the ones in the gut, where receptors for sweet and umami (and perhaps fatty acids, as well) signal to the brain that a nutritious meal has arrived. This helps us learn what flavors we should seek when we're looking for our next meal. Our guts have bitter taste receptors, too, which may activate defensive responses to toxins. A few researchers have suggested that these may be responsible for some of the side effects of bitter-tasting medicines.

We even have bitter receptors in our respiratory passages,

of all places. Why do we need to taste the air we breathe? Well, because it has bacteria. As it turns out, one of the chemicals that bacteria use to communicate with one another has a bitter taste. Bitter receptors in our sinuses and the lining of our bronchial passages detect this and alert our immune system to fight back against the invaders. Curiously, the bitter receptor responsible for this, *T2R38*, is the same receptor that determines our sensitivity to PROP and phenylthiocarbamide (PTC). And, in fact, people who can't taste PROP—that is, who have a broken *T2R38* receptor—turn out to have more sinus infections. Some researchers even think that bitter receptors may have evolved originally as part of the immune-defense systems of our ancient animal ancestors, and only later turned out to be useful in our mouths, too. If so, we have disease to thank for much of the flavor of coffee, beer, and broccoli.

By now, you might have noticed a glaring gap in our taste repertoire. The sense of taste is all about identifying good stuff to put in our mouths: sweet carbohydrates, salty sodium, protein-rich umami. It also helps us recognize bad stuff we want to avoid eating, such as sour, unripe fruit and bitter, poisonous plants. But there's another category of good stuff we haven't talked about yet, one that might be the most treasured of all: fat. Surely, our taste system ought to have evolved to recognize this energy-rich and (in the prehistoric world of our ancestors) scarce resource. And in fact, we probably can. In the past few years, researchers have been piling up a convincing heap of evidence that suggests we ought to include a sixth taste, fat, in addition to the familiar five. But there's a surprising twist to the story: we hate the taste.

Rick Mattes, a nutrition scientist at Purdue University in Indiana, probably knows more about our taste for fat than anyone in the world. The fats we find so appealing in foods—the butter we put on our bread, the olive oil in our salad, the cream on our strawberry shortcake—are what chemists call triglycerides. These are big molecules composed of a backbone molecule with three so-called fatty acids stuck to it, like a small box kite with three long tails. There's no evidence that triglycerides have any taste at all, says Mattes. Instead, we recognize them in our mouths through the sense of touch, which picks up their creamy lubricity.

On the other hand, there's more and more evidence—much of it from Mattes and his colleagues—that we do indeed taste fatty acids when they become separated from their backbone. We have receptors on our taste buds that recognize fatty acids and respond by sending electrical signals to the brain's taste center.

And the taste seems to be distinct from any of the other five primary tastes. That's easy to show in rodents, by pairing a nausea-inducing chemical with a fatty acid taste. The rats quickly learn to avoid the sick-linked taste, just like a hangover from too much rum and Coke can put you off of cola for a while. But the fat-avoiding rats don't avoid sweet, sour, salty, bitter, or umami tastes, which implies that their learned aversion is to a sixth taste instead. Mattes has shown that humans, too, perceive fatty acids as a distinct taste. Since "fatty" calls to mind an oily texture, rather than a taste, Mattes suggests using the term *oleogustus* (Latin for "fatty taste") for the taste sensation.

By this point you might be wondering: If fatty acids have their own taste, what is it? Not good, it turns out. "They are really awful," says Mattes. Most of the time, free fatty acids—that is, ones that aren't bound up as triglycerides—signal decay or rancidity. In fact, the food-processing industry spends a lot of time and money trying to keep free fatty acids below detectable levels in their products. If you want to know what free fatty acids taste like, says Mattes, find a batch of old french-fry oil that's gone rancid. Hold your nose, to eliminate the strong odor, and then taste it. But don't expect to be able to describe it. "If you ask people to give you a description of it, it's like they have blinders on," says Mattes. "We don't have a language for it. They'll frequently call it bitter or sour, but what I think they mean is that they don't like it."

So it looks like our ability to taste fatty acids is more like our taste for sour or bitter—that is, a defense against eating the wrong things—than like our taste for sweet, salty, or umami, all of which signal the right things to eat. But the story might be a little more complex than that, Mattes thinks. After all, we know of other cases where a tiny bit of an unpleasant taste actually enhances the overall flavor of a food. "Wine without a little bit of bitterness would not be as good," he says. "Chocolate without bitterness would not be as good." In the same way, a hint of the nastiness of fatty acid taste does pop up in a few foods we learn to like, most notably some fermented foods and stinky cheeses.

As the evidence for fat taste piles up, more and more experts are now willing to add it to the list, expanding our repertoire of basic tastes from five to six. And there might be other basic tastes out there, too. There's some evidence that we have a taste for calcium, and for carbon dioxide. Rodents look like they have a taste for starch, though it's not clear that humans do as well. Some researchers even suggest that we might have a basic taste for water. And there's a mysterious one called *kokumi*, which many Asian researchers think might qualify as yet another basic taste—though many Western scientists remain skeptical. Weirdly, *kokumi* seems to have no taste of its own, but when you add it to something that already has a salty or umami taste, it enhances those flavors.

In a lab at Monell, I tasted some popcorn sprinkled with *kokumi* powder. It had a haunting, elusive flavor that was hard to put my finger on—sort of cheesy, sort of meaty, like the flavor powder on the surface of Doritos. Clearly, *kokumi* does something to taste perception, but it's hard to say exactly what. (You can taste it yourself—look for *kokumi* powder at Korean groceries.) Scientists don't know exactly how we perceive *kokumi*, though a calcium-sensing receptor called (surprise!) the calcium-sensing receptor seems to be involved. Things are changing fast in this field. Who would have thought there could be such complexity to something as seemingly simple and obvious as the four basic tastes we learned in school?

To complicate things still further, the basic tastes interact with one another. Salt suppresses our perception of bitterness, as we've seen. Similarly, sweet and bitter suppress each other. Tonic water is a great example of this: The bitterness means we don't notice how sweet the drink actually is, while the sugar helps bring the bitterness down to a level that most of us find palatable. Except for people like Linda Bartoshuk, of course.

Which brings us back to supertasters. The ability to taste PROP turns out to be mostly a function of one particular bitter receptor, *T2R38*. There are two common variants of this gene: one version that responds strongly to PROP and one that doesn't. This suggests that people with two copies of the nonresponding gene (one from each parent) are nontasters, those with two copies of the high-responding gene are supertasters, and those with one of each are normal tasters. And, indeed, researchers sometimes genotype people for *T2R38* as a quick, objective way to determine their taster status.

But it's not that simple. The *T2R38* receptor recognizes just one group of chemicals: those that contain a particular set of atoms called a thiourea group. Your ability to taste those should have nothing to do with your ability to taste sweet, salty, or other kinds of bitter—including quinine—let alone your perception of the burn of chili peppers, which involves an entirely different set of receptors and nerves. And it certainly shouldn't affect the number of fungiform papillae on your tongue.

T2R38 probably has nothing to do with supertasting, at least not directly. Your *T2R38* genes determine whether you have the genetic ability to taste PROP at all—but if you do, the amount of bitterness you experience probably depends on how well the rest of the taste machinery in your mouth and brain responds. The genes that control that machinery are what really make the difference between a taster and a supertaster—and if you can taste PROP at all, the amount of bitterness you experience is a decent measure of how sensitive the rest of your machinery is. That's probably why people who rate PROP as intensely bitter also tend to rate salt as saltier, sugar as sweeter, and chili peppers as hotter than people who find PROP less bitter. If so, people with broken *T2R38* genes might still be supertasters for anything that doesn't require that bitter receptor. They just need to find a different way to prove it.

One way might be to measure the density of fungiform papillae, which is why Bartoshuk painted my tongue blue. Each papilla contains several smaller clusters of cells bearing taste receptors. These clusters are the real taste buds, technically speaking, and the cells within them send nerve impulses up the taste nerves to the brain, signaling which of their receptors has encountered its particular taste quality. It makes sense that tongues with more papillae would generate stronger nerve signals and hence experience more intense tastes. Sure enough, most studies do support that hunch—although there are a few annoying studies that fail to find a link between number of papillae and taste perceptions.

So what determines how many papillae you have on your tongue? Nobody knows for sure, but there are intriguing hints that a protein called gustin might be involved in stimulating the formation of fungiform papillae. People with one particular variant of the gustin gene have abundant, normal papillae, while those with a different variant have large, misshapen, sparsely scattered papillae. No doubt, too, there are plenty of other genes that affect overall taste sensitivity and thus help to define whether you're a supertaster, an ordinary taster, or a (relative) nontaster. But the science doesn't seem to have caught up with our curiosity on this matter.

Fortunately, scientists do know a fair bit about the genetics that underlie some of the differences in people's taste perceptions—enough, in fact, to make it clear that each of us lives in a unique world of flavor. Genetic differences likely explain some (though not all) of why former president George H. W. Bush hated broccoli, why a gin and tonic is ambrosia to one person and anathema to another, or why some of us put sugar in our coffee. I wanted to learn more—and, especially, I wanted to know where my own taste perceptions fit into the picture. Once again, that brought me to Monell.

In particular, I wanted to see Danielle Reed, who has done a lot of the best work on genetic differences in taste perception. A few months before my visit, I had drooled into a vial and shipped it off to Reed for genetic analysis. (Saliva contains enough cells that geneticists no longer need blood samples or even cheek swabs to run their DNA tests.) Now it's time to see how my sense of taste compares with everyone else's.

Reed's taste-test procedure couldn't be more low-tech. Her assistants hand me a box containing several numbered vials of liquid, plus a large plastic cup to spit into. Starting with vial 1, I sip the liquid, swish it around in my mouth, and spit into the cup, indicating on a questionnaire how sweet, salty, sour, and bitter I found the sample; how intense the sensation is; and how much I like it. And then I go on to vial 2. It's a bit like a wine tasting, but without the pretentiousness. And without the wine.

A few hours later, test scores in hand, it's time to sit down with Reed to see how they match up with my genes. In person, Reed is a short, plump, cheerful woman with frizzy dark hair who clearly thinks unpacking someone's genes is a bit like unwrapping a present. She must have done this hundreds of times by now, if not thousands, but the excitement is still there.

The first test turned out to be a bit of a trick: Vial 1 held plain old distilled water. I'm relieved to see that I scored its taste intensity to be "like water"; rated it dead neutral on the liking spectrum; and detected no sweet, salty, sour, or bitter tastes. At least I'm not tasting stuff that isn't there. Now on to the real tastes, and the genes.

First up, *T1R3*, the gene that contributes to the receptors for sweet and umami. Reed had tested my genome for a variant that, other researchers had found, affects sweet perception. These genetic variants are like spelling changes in the genome. Just as changing a single letter—"dog" to "dig," say—can alter a word's meaning, changing a single letter in the DNA sequence of a gene can alter the resulting receptor protein. For the *T1R3* variant, people with a *T* at one particular spot are less sensitive to sweet taste, and like it more, than those with a *C*. "It's like they can't taste sweet as much, so they are choosing the higher concentrations," says Reed.

I turn out to be a *TT*—one *T* from each parent—which should make me a classic sweet craver. But that really didn't make sense, I told her. Just that morning, I'd been given a sweetened iced coffee at Starbucks by mistake, and I had ended up pouring most of it out, because it was much too sweet to drink. As far as I'm concerned, it's also no big deal to skip dessert after dinner—it's not important to me. Had something gone wrong with the genotyping?

Reed turned to my taste-test result and burst out laughing. "Oh, look at you! You're not so far off here." I'd rated the 12 percent sugar solution—roughly equivalent to a (flat) soda—as only moderately sweet, and highly pleasant. Reed herself—a *CC*—finds it disgustingly syrupy. Clearly, the link among genes, taste perceptions, and actual food choices is not a simple one.

That complexity is also evident in some of my bitter receptor genes that Reed tested. One of these was the bitter receptor *T2R19*, which detects quinine, the bitter chemical found in tonic water. I had the low-responding gene variant, according to the genetic test. Sure enough, when I sipped Reed's quinine solution, I scored it only mildly bitter and not very intense. That squares nicely with my liking for tonic water, which you may recall is about the only soft drink I ever drink. But it doesn't explain Reed's fondness for gin and tonic, because she carries the high-intensity gene variants. "I taste gin and tonic as very bitter," she says, "but I love it!"

Then there's our old friend *T2R38*, the bitter receptor that determines sensitivity to PROP, PTC, and the bitter thiourea compounds in broccoli and brussels sprouts. The genetic test backs up what I already knew from talking with Bartoshuk: I'm one of the "lucky" ones who reacts strongly to these bitter chemicals. And when I tasted the PTC solution, I scored it as intensely bitter.

So why does Dani Reed like gin and tonic, which she finds intensely bitter? Why am I drawn to the foods and drinks I taste as bitter, instead of avoiding them?

"What you taste isn't always what you like," says Reed. "I always say, 'It's the brain, stupid!' You can learn! Within the correct context, it's very much beloved." Indeed, we quickly learn to find pleasure in flavors—even ones we initially find repulsive—that are paired with attractive rewards. The bitter coffee that delivers a wake-up jolt soon becomes pleasant in its own right. Same for the bitter beer or gin and tonic that accompanies an evening with good friends.

There may be another dimension to taste preferences, too, says Beverly Tepper, a sensory scientist at Rutgers University in New Jersey. Some of us are what Tepper likes to call "food adventurous." That means there are really two kinds of supertasters, according to Tepper. Those who are not food adventurous are the classic, picky eaters: they don't like things too sweet, too hot, too fatty, too spicy. "They know what they like, and their food choices are guided by their previous experiences. They're a little bit finicky," says Tepper. Mr. Lima-beans-and-milk presumably falls into that category.

On the other hand, supertasters who are food adventurous are willing to be surprised, even by intense tastes, and will try something again even after a disconcerting first experience. Because they're not put off by intense experiences, this category of supertasters resembles nontasters in their food preferences.

"I'm a supertaster, and I actually like a lot of the foods that theoretically I shouldn't like. But I'm also food-adventurous," says Tepper. That describes me to a *T*, too. I get the intense sensory jolt from a highly flavored food—but I like the stimulation.

These few genes that I had tested are probably just the tip of the iceberg when it comes to genetic differences in taste perception. Reed thinks there could be dozens—perhaps even hundreds—of genes that affect our taste acuity and our perceptions of particular tastes. In addition to the taste receptor genes themselves, many other genes probably affect how our cells respond once a taste receptor has been stimulated, how readily signals are sent to the brain, and every other step of the taste-sensing pathway. My flavor world, it seems clear, is different from yours. We can serve ourselves from the same bowl of soup and have different taste experiences. And taste is only one part of the flavor equation.

Chapter 2

BEER FROM THE BOTTLE

The Association for Chemoreception Sciences, North America's main conference for smell and taste researchers, meets every April in southern Florida. The location isn't accidental—the whole point is to give researchers the opportunity to leaven their scientific geekery with at least a few hours of sun and sand. This lends the meeting a remarkably relaxed, nonacademic feel, with sun-deprived, middle-aged folks clad in shorts and Hawaiian shirts thronging the bar or basking poolside in the sun. But that stereotypical Florida hotel ambiance quickly turns surreal, as the conversations on the sundeck turn not to shopping or the kids, but to G-protein-coupled receptors, the psychophysics of odor perception, or the olfactory abilities of mosquitoes. For four days in April, the Hyatt Regency Coconut Point in Bonita Springs is not your average Florida resort.

When attendees are not by the pool or talking science in the bar, they can often be found in the exhibit hall, where they can peruse posters that describe current research, or browse new scientific gadgets that vendors are selling. That's where I first met Richard

Doty, who was looking relaxed and informal in a green and black-striped rugby shirt. Doty—a fit-looking seventy-year-old with short, gray-tinged hair and a cheerful manner—is one of the world's leading experts on the senses of smell and taste. In fact, he literally wrote the book on the subject: His *Handbook of Olfaction and Gustation* is the classic in the field. But even if you didn't know that, you could guess his stature by the steady stream of eminent scientists who stop by to chat. Right now, though, Doty is playing the role of pitchman. The company he founded is hawking a new machine for testing people's sense of smell, and they're inviting all comers to try it out. Clearly, that's an opportunity I can't pass up.

Specifically, Doty's machine is designed to measure olfactory threshold, an indication of how sensitive your sense of smell is. By "olfactory threshold," he means the most diluted trace of an odorant you're able to detect; the lower your threshold, the more acute your nose. One of Doty's assistants took me through the process. You sit in front of the machine and put your nose into this little mask, he explained. Then the machine will give you two puffs of air, one after the other, and the computer will ask you which of the two carried the scent of phenylethyl alcohol, a pleasant roselike odor. And then you repeat the test again and again, until the computer instructs you to stop.

What the assistant didn't tell me—but Doty did later—was that the olfactometer could vary the concentration of the rose scent in the loaded puff. If I failed to answer correctly which puff had the scent, the computer assumed there was too little scent for me to detect, and it stepped up the dose for the next round; if I answered correctly, it assumed the concentration was above my detection threshold and reduced the dose. Over and over we went, wandering up and down like a hyperactive kid on a staircase, until we settled on the odor concentration that sat at the boundary between right and wrong answers—my olfactory threshold.

At this point, Doty strolled over and glanced idly at the printout of my result. His eyebrows went up. He stopped and peered more intently at the printout, and then turned to me with an expression of concern. "Do you have an impaired sense of smell?"

Uh-oh. When the world expert on olfactory dysfunction takes an interest in my test results, that can't be a good sign. Especially for me, especially now: How can a guy with an impaired sense of smell credibly write a book about flavor? (You'll recall from the jelly bean test that flavor is mostly about the sense of smell.) As Doty showed me the printout, the news looked pretty grim: according to his machine, the rose scent had to be present in more than one part per thousand before I could reliably detect it, making my threshold about a thousand times worse than average.

Doty must have seen the pained expression on my face, because he pulled an envelope from a nearby box and said, "Here, why don't you take this test, too?" The envelope contained another of Doty's many claims to fame, the University of Pennsylvania Smell Identification Test. This test, universally referred to as the UPSIT, is a forty-item multiple-choice test that uses scratch-and-sniff scents. ("This odor smells most like a. gasoline b. pizza c. peanuts d. lilac." *Pizza*, I thought.) Picking one of four multiple-choice answers avoids the well-known difficulty people have in putting a name to a smell. Most of the time, the right answer seemed obvious, but maybe five or ten of the forty were tough. "Is this turpentine or Cheddar cheese? I'm not sure," I found myself saying. Even distinctive smells can be hard to recognize sometimes.

A few hours later, I bumped into Doty on the exhibit floor again and gave him my UPSIT for scoring. To my relief, I got thirty-seven of the forty right—enough to put me in the seventy-third percentile for fifty-five-year-old men. "You did very well," said Doty. "Three-quarters of your friends did worse." Whew! My nose doesn't disqualify me after all.

Most likely, Doty speculated, the problem with the threshold test was the environment we were in: A bustling exhibit hall isn't the ideal place to concentrate on subtle, barely detectable odors. Plus, I'd raced through the test as quickly as I could so that the next person could try it; in a doctor's office, the test is given much more slowly, with pauses that allow the scent from one trial to dissipate fully before the next trial starts. These minor differences in procedure can make a huge difference to the outcome—a complication that colors almost all research on the sense of smell.

That was my introduction to the messy world of olfaction research, where everything is harder—and more complicated—than it looks. While taste research is enjoying something of a golden age, smell researchers are, for the most part, still mired in the Dark Ages. Given an unknown molecule, even the best scientists have only recently been able to predict whether it has an odor at all, and can barely guess at what that odor might be. In fact, researchers can't even agree on the details of how olfactory cells recognize odor molecules. All of which means that we're a long way from understanding the most important mystery of the sense of smell, at least from the perspective of flavor: Do your perceptions differ from mine, and if so, what does that mean for our appreciation of flavor?

The reason olfaction has proven such a tough nut to crack is that it's much, much more complex than taste. As we saw in the last chapter, these two flavor senses really serve two different purposes. Taste draws us toward nutritive foods and pushes us away from poisonous ones—a fairly simple yes/no decision. That makes taste the easy part of the flavor equation: Our tongues use at most thirty or forty receptors to keep track of a half-dozen or so basic tastes. It's pretty straightforward to understand what we're talking about, and how our sense of taste works. Smell, on the other hand, answers the question "What is it?" which is a much more open-ended question. There are, after all, a vast number of smelly things out there in the world, and our noses need to be able to cope with all of them.

Imagine taking a whiff of your morning coffee. The steam rising from your cup carries with it hundreds of different aromatic molecules, which enter your nose as you sniff. Way up at the top of your nasal cavity is a little patch of cells, less than one square inch in area, called the olfactory epithelium. The nerve cells within this patch—about six million of them—each carry one of about four hundred different odor receptors on their surface. (Actually, a few cells major in one receptor and minor in another, but we can ignore that detail here.) These olfactory sensory nerve cells send their signals straight in to the brain, giving them the distinction of being the only nerve cells in your body that connect the brain directly to the outside world.

Each receptor, in turn, recognizes particular features of specific odor molecules from the coffee. Surprisingly, scientists still don't know for sure how this recognition happens. Most think that particular shapes on the odorant molecules fit into complementary shapes on the receptors, like the camera-in-foam-case analogy we used for bitter receptors. A vocal minority, however, thinks that instead, each odor molecule has a unique pattern of molecular vibrations, which receptors recognize using an arcane process called quantum tunneling. A lively debate is still raging between the "shapists" and the "vibrationists," though of late it looks like the shapists are winning.

For most purposes, though, it doesn't matter exactly how this recognition happens. What's important is that each odor receptor recognizes several to many different odorants, and each odorant binds to several different receptors. That means that each odor molecule activates a different mix of receptors—a different chord, if you will, on the olfactory keyboard. And your coffee contains not just one odor molecule but hundreds, each sounding its own distinctive chord in your brain. Some of those chords probably sound so faintly that you can't actually "hear" them as part of your flavor experience. (In technical terms, their concentration is below your detection threshold.) But that still leaves a whole orchestra's worth of important chords, as each above-threshold odorant tickles its own particular mix of receptors. Out of that cacophony, your brain somehow extracts a harmony: the flavor you know as coffee.

No wonder olfaction is so hard to understand. It has three separate sorts of complexity: diverse odor molecules, diverse receptors, and diverse "harmonies." Let's look at each one in turn, starting with the molecules. No one knows exactly how many different odor molecules there are in the world. For many decades, the standard answer to that question has been "about 10,000." You'll see that number bandied about everywhere from chefs' blogs to scientific papers to neuroscience textbooks. Even Richard Axel and Linda Buck, who won the Nobel Prize for finding the receptors responsible for detecting odors, used it in their key paper. Bathed in Nobel glory, the notion of 10,000 different odors has come to take on the aura of received wisdom. And it adds to our general sense of incompetence when it comes to the human sense of smell. After all, psychologists estimate that we can recognize as many as 7.5 million different colors and 340,000 audible tones. Compared with that, recognizing 10,000 smells is pretty pathetic.

But a closer look shows that this 10,000-smells number, far from being hard science, is completely bogus. It comes from a seat-of-the-pants calculation dating way back to 1927. Two chemists, E. C. Crocker and L. F. Henderson, thought that smells, just like tastes, could be sorted according to four independent qualities. For taste, we have sweet, sour, salty, and bitter. (We can cut them some slack for missing umami, which few except the Japanese knew about back then.) For smell, they suggested fragrant, acid, burnt, and one more, which they first called putrid and later changed to caprylic, or goaty. And they further guesstimated that each of the four odor qualities could be assigned an intensity score between 0 (absent) and 8 (overwhelming). If so, there are $9 \times 9 \times 9 \times 9$ different ways to score a smell, a total of 6,561, which they generously rounded up to 10,000. Of such stuff is scientific orthodoxy made. If Crocker and Henderson had chosen to include a fifth quality—musky, say—and rate on a scale of 0–9, we would all have been talking about a universe of 100,000 smells instead.

So far, so bad. Joel Mainland, an olfaction researcher at Monell, thinks he can do better. Mainland is a compact, enthusiastic guy with a thin face, wire-framed glasses, and rapid speech. He started out in science thinking he would study vision, but realized early on that it would be hard to build a career there. "As I looked around the field, I realized that the big problems were solved," he says. "And then you look at olfaction and the big problems are still not solved. To me, it was an easy switch to go to olfaction." His hunch has paid off in spades: Mainland has become one of the brightest rising stars of olfaction research.

Recently, Mainland has tried to come up with a more educated guess at how many different odor compounds there are in the world. His reasoning goes like this: In order for us to smell a molecule, it has to be volatile—in other words, willing to launch itself into the air in gaseous form. Big molecules generally can't do that, and in fact, chemists know of few smelly molecules that have more than twenty-one "heavy" atoms in them—that is, atoms other than hydrogen, the atomic featherweight. So let's assume, he says, that only molecules with twenty-one or fewer heavy atoms could have odors. That gives us, by his estimate, about 2.7 trillion candidate molecules.

But not every one of those small molecules actually has a scent. Some have boiling points so high that they never become airborne at normal temperatures; others are so oily that they're repelled by the watery mucus layer that lines the nose, so they can't activate odor receptors. After some tinkering, Mainland and his colleagues came up with a way to use a molecule's oiliness and boiling point to predict whether it would be smelly.

One morning in Mainland's lab at Monell, I helped test some of his predictions. It turns out you can't just give someone a sample and say, "Do you smell anything?"—the power of suggestion is so strong that they'll often "notice" an odor that's not really there, or pick up some stray odor in the room. Instead, the researchers use something called a "triangle test." Mainland's assistant sat me down at a table and blindfolded me, then waved three vials under my nose, one at a time, as a synthesized computer voice asked which one—A, B, or C—had the odor. After each set of three, they gave me a thirty-second "distraction break" to avoid nose fatigue: the computer played a short song clip and asked me whether the singer was male or female. (Mainland had intentionally picked ambiguous voices, so this was hard. Showing my age, I got Tiny

Tim and a young Michael Jackson right, but was clueless on much of the contemporary stuff.)

Tests like these, performed on many different individuals, give Mainland the confidence to say that most people have a hard time telling male singers from female ones. More to the point, he also knows he's about 72 percent correct in predicting whether an unknown molecule will have an odor. Applying his prediction method to the whole universe of 2.7 trillion candidates, he calculates that there must be a staggering 27 billion different smelly molecules in the world.

That's not the same thing as saying there are twenty-seven billion different smells, though. After all, we know that several different molecules have an apparently identical sweet taste, and there might be hundreds of different molecules that give rise to a single bitter taste. If the odor universe is similarly full of "smell alikes," then the number of unique odors could be much, much less than twenty-seven billion. But when I asked Mainland if he knew of any two molecules that smell exactly alike, he couldn't think of any. "I was always told that no two molecules smell the same," he said.

Now let's switch over to the other side of the equation and look at the receptors that are responsible for detecting all those smelly molecules. Buck and Axel showed that the odor receptors are protein molecules embedded in the membranes of nerve cells in the olfactory epithelium. When geneticists first sequenced the human genome a few years after Buck and Axel's discovery, they therefore knew an odor receptor gene when they saw it. To their astonishment, they found not just a few dozen olfactory receptor genes in the genome, but nearly a thousand! Stop and think about that for a moment: The human genome contains about twenty thousand genes in all, so out of all the genetic instructions needed to turn a fertilized egg into a functioning human being—hundreds of cell types organized into tissues and organ systems and a brain, all the molecular signals needed to keep everything running—one out of every twenty genes is for an odor receptor. That's like walking into a library containing the world's accumulated knowledge and finding that one in twenty books is about car repair. Who would have guessed that olfaction makes up such a large chunk of who we are?

On closer inspection, more than half of these odor receptor genes turned out to be what geneticists call "pseudogenes"—that is, the rusted-out hulks of genes that had broken sometime in our evolutionary past. Exactly how many odor receptor genes are still functional is a bit tricky to answer. The official human genome—largely that of the flamboyant genetic entrepreneur Craig Venter—has about 350 working odor receptors. But if the Human Genome Project's gene sequencers had looked instead at your genome, they would have found that some of those 350 are broken in your genome, while others that were broken in the official version are working in yours. One team of researchers looked at a sample of one thousand human genomes and found 413 odor receptors that were functional in at least 5 percent of the population. If the researchers had looked at more people, they would no doubt have found a few more.

It's one thing to count odor receptor genes, though, and quite another to understand which receptors recognize which odor molecules. The latter is much harder, largely because odor receptors normally live on the surface of nerve cells, which are challenging to grow in petri dishes in the lab. That makes experimentation difficult. As a result, the vast majority of receptors are what scientists, in a rare burst of colorful metaphor, call "orphan" receptors, meaning that we don't yet know which odorant molecules they recognize.

Fortunately, molecular biologists have found a work-around by putting odor receptors onto the surface of kidney cells, which are much easier to grow in the lab. A few years ago, with a bit of hard work, Mainland and other researchers created a panel of kidney cell cultures expressing the whole range of human odor receptors, one per culture. With the panel in place, they looked forward to testing odorants, one after another, to see which receptors they triggered. Soon, they thought, they'd be able to "de-orphan" the lot. The olfactory code looked within reach at last.

No such luck. So far, Mainland and the other workers have only managed to find targets for about 50 human odor receptors. Try as they might, the other 350-odd receptors have remained stubbornly orphaned. "That means that about 85 percent of these receptors do not work in our assay system," says Mainland. "That's a lot." It's possible that the apparent failures detect uncommon odorants that Mainland simply hasn't got around to testing yet—though the longer he looks, the less likely that possibility becomes. It's also possible that some overlooked complication is preventing those receptors from working properly in the kidney cells.

There's another, more interesting possibility: Maybe some of our odor receptors aren't there to detect odors at all. If you take a step back and look at the big picture, what odor receptors really do is to alert the body when they recognize particular small molecules in the environment. Some of those molecules are odors, but this sort of recognition plays lots of other roles, too. Our bodies need to recognize hormones and other signaling molecules that help the body keep organized during growth and development; they need to turn functions like digestion, reproduction, and immune defense on and off at the right times, and so on. Since evolution is the ultimate MacGyver, cobbling together solutions from whatever materials happen to be lying around, it would be surprising if at least a few odor receptors hadn't been pressed into service for other functions now and then. Sure enough, when biologists have looked, they've found ORs all over the place: testis, prostate, breast, placenta, muscles, kidneys, brain, gut, and more. Some of these, no doubt, occur in the nose as well—but it's at least possible that some do not.

But counting up odor receptors doesn't tell the whole story of smell, because there's another whole layer to the way we perceive odors that isn't there for taste. Our sense of taste is what sensory scientists call analytic—that is, we easily break it down into its component parts. Sweet and sour pork is, well, sweet and sour. Soy sauce is salty and umami. Ketchup is sweet, sour, salty, and umami.

Our sense of smell doesn't work that way. Instead, it's a synthetic sense: Our brains assemble the component parts into a single, unified perception, and we can't easily pick out the parts separately. That's easiest to understand if you think about another synthetic sense: vision. When I gaze fondly at my wife, I don't see lines, curves, and edges, even though that's what my brain is actually detecting and processing. I just see her face, the synthetic object of my perception. Similarly, the individual odor molecules sensed by our nose can combine in our brain to create a new perception that's entirely different from its components. If you combine ethyl isobutyrate (a fruity odor), ethyl maltol (caramel-like), and allyl alpha-ionone (violetlike) in the proper proportions, for example, what you smell is not caramel-coated fruit on a bed of violets, but pineapple. Similarly, one part geraniumy 1,5-octadien-3-one to one hundred parts baked-potatoey methional smells fishy—something neither ingredient shows the least hint of alone.

Neuroscientists like to refer to these new, higher-level perceptions as "odor objects." Each one is, in effect, a unique pattern of activation involving a subset of the four hundred or so different kinds of odor receptors in your nose. In essence, these odor objects define reality in our olfactory worlds, just like my wife's face is a visual object that seems more real to me than its component lines and curves.

And in the same way that you can create an essentially infinite number of faces out of a smallish set of lines and curves, our four hundred odor receptors can give rise to a dizzying number of different odor objects. A few years ago, researchers gave people mixtures of ten to thirty different odor molecules and asked whether they could tell them apart. Based on those results, they calculated that people ought to be able to distinguish at least a trillion different odor objects—a big step up from the fabled ten thousand smells of received wisdom. (By comparison, sensory scientists say our eyes can perceive a few million different colors and our ears maybe half a million pitches.) Since then, other researchers have pointed out that the "one trillion" number should be treated with caution, since it depends on several iffy assumptions. However, the general message—that the universe of smells is a huge one—still stands.

To understand how the brain processes these odor objects, I sought out Gordon Shepherd, one of the grand old men of olfaction research. Nearly everyone I spoke to at the Association for Chemoreception Sciences meeting in Florida, in fact, made a point of saying, "You should talk to Gordon Shepherd." Some even suggested that he's been so important to research on the neuroscience of smell that he deserved a share of the Nobel Prize for his work. He's also written a terrific book, *Neurogastronomy*, about the biology of flavor perception.

When I caught up with Shepherd on the resort patio, I found a courtly, white-haired man in a red wool sweater, who was happy to spend the afternoon talking about olfaction. Odor objects have a physical equivalent in the brain, he told me. Each one of the nose's four hundred odor receptors delivers its signal to a different part (or parts) of the brain's olfactory bulb, the first relay station for odor information. If you imagine the olfactory bulb as a switchboard panel with lights corresponding to individual odor receptor types, then each odor object is represented by a distinct pattern of lights—its own olfactory image, in effect. But when your brain comes to process that pattern of lights, it doesn't know whether they're the result of a single odorant molecule or many: it just sees the pattern.

And we're generally very bad at articulating complex patterns, says Shepherd. Just try to describe the face of someone familiar to you, or the art of Cy Twombly—you'll probably struggle just as much as most people do in expressing the aroma of a beefsteak tomato or an artichoke. "It's the same problem," says Shepherd. "A highly complex image that's almost impossible to describe in words."

That certainly matches most people's experience of talking about smells—and by extension, about flavor. Putting names to smells is something humans in general are "astonishingly bad at," says Noam Sobel of the Weizmann Institute of Science in Israel, one of the most creative, and consistently provocative, smell researchers. To prove this incompetence to a skeptical family member, Sobel once asked her to close her eyes, then he pulled a jar of peanut butter out of the fridge, removed the lid, and waved it under her nose. Even though his relative ate peanut butter almost every day of her life, she couldn't name that familiar smell. You can repeat the test yourself: Close your eyes and have a friend present you with some familiar household odors, and see how many you can identify. You'll probably find, as Sobel and other researchers did, that you recognize all of them as familiar, but you can't name even half of them successfully. (I once failed to identify the flavor of coffee, which at the time was my every-morning breakfast drink.) As one of Sobel's colleagues is fond of pointing out, if you or I did that badly at naming colors or shapes, we'd go straight to a neurologist to see what's wrong.

Another big reason we're so bad at naming smells is that our brains process odor information—one of our most ancient senses—much differently than they handle newer senses like sight and hearing. Sights and sounds take an express route to the thalamus, the part of the brain that acts as the gatekeeper of consciousness. We're wired to pay conscious attention to them. That direct line also means that sights and sounds have rapid access to the newer, more powerful brain regions that handle speech and language. In contrast, olfactory signals go first to the ancient, preconscious brain regions that control emotion and memory, the amygdala and hippocampus—which helps explain why smells are so powerfully evocative—and don't pass through the gateway to consciousness and language until several stops later.

But there's a second reason for our difficulty. In English—and most other Western languages—we pretty much lack a distinct vocabulary for describing odors. We describe smells, if we can describe them at all, by saying what they're *like*: a New Zealand sauvignon blanc smells grassy, we say, or a furniture polish smells lemony, and that's about the best we can do. Here's an English-speaking American trying to put a name to the smell of cinnamon: "I don't know how to say that, sweet, yeah; I have tasted that gum like Big Red or something tastes like, what do I want to say? I can't get the word. Jesus, it's like that gum smell like something like Big Red. Can I say that? Okay. Big Red. Big Red gum." You've probably flailed about in a similar way trying to describe a smell—I certainly have. But we don't do that for colors, for which we do have a specialized vocabulary. We don't have to describe the colors of the Swedish flag, say, as lemonlike and skylike—we can call them yellow and blue.

And as it turns out, some cultures do that for smells, too. For a startling example of just how much better we could be at recognizing, identifying, and talking about smells, consider the Jahai, a small tribe of nomadic hunter-gatherers in the mountains of southern Thailand, near its border with Malaysia. The Jahai language has more than a dozen words to describe smells, none of which relate to the smell of any particular object. The Jahai might say, in their language, that something smells "edible" or "fragrant," or, my favorite, "attractive to tigers." Some of the actual concepts they're expressing make no sense at all to an outsider—the word for "edible," which sounds a bit like "knus," is applied to gasoline, smoke, bat droppings, some millipedes, and the wood of wild mango trees, none of which strike me as particularly edible; "fragrant" includes several flowers and fruit, some other kinds of wood, and a species of civet cat.

However bizarre it seems to us, that specialized vocabulary makes it much easier for the Jahai to think and talk about smells. When researchers gave a standard smell-identification test to ten Jahai men, they found that the Jahai tended to be quick and consistent in describing the smells, even though most of the actual odors used in the test were unfamiliar to them. In fact, the Jahai proved to be just as comfortable describing odors as they were at describing colors. By comparison, ten English-speaking Texans were quick and precise at describing colors, but all over the map when it came to the smells. (One of those Texans was the source of the hopelessly inarticulate description of cinnamon quoted above.)

Fortunately, vocabulary is something we can learn with a little effort. Even we Westerners have specialized vocabularies for smells within certain domains. Just listen, for example, to a professional perfumer pick apart the olfactory spectrum of a fragrance, nimbly identifying the floral top notes, musky base notes, and the like. An experienced wine buff can do the same thing with what's in her glass. In fact, tests show that wine experts' noses are no better than yours or mine—they've just had more practice at noticing and putting into words what they're smelling. Almost anyone can improve their nose for wine, no matter how hopeless they feel. As long as you can recognize that one wine is different from another, you've got the basic perceptual tools you need. All it takes is a little effort to nail down the vocabulary.

But there's a limit to how well even the professionals can deconstruct the aromas of a glass of wine or a whiff of perfume. Way back in the 1980s, Australian psychologist David Laing presented volunteers with familiar odors like cloves, spearmint, orange, and almond singly or in combinations of up to five. He provided a list of seven possible odors and asked the volunteers to check off all the odors that were present. People did okay at single odors or mixtures of two, but their performance fell off dramatically at three or more. Not a single person correctly identified all the elements of a five-odor mixture. Later studies have confirmed this result—even professional flavorists and perfumers just don't seem to be able to correctly identify more than three or four odors in a mixture, probably at least partly because the odors interfere with one another in our nose or brain. With this in mind, I'm inclined to look skeptically at wine tasting notes that claim to identify six or eight aromas.

Are there ways to help identify odors? That is, can we somehow sort odors into categories to make it easier to understand them? We do that for taste, after all: There's sweet, sour, salty, bitter, umami, and maybe a few others. Color and sound are also simple to sort: It's all about the wavelength of light or the frequency of a sound vibration. But odors are caused by thousands to billions of unique molecules, each with a different shape and each, apparently, activating a different set of odor receptors. How to make sense of all this?

Of course, people have tried, beginning long before they knew anything about molecules. Carl Linnaeus, most famous for his method of classifying all living beings, had a go at odors, too. All odors, he thought, fall into seven categories: fragrant, spicy, musky, garlicky, goaty, repulsive, and nauseating. A contemporary, Albrecht von Haller, had an even simpler system, sorting all odors somewhere on a spectrum between ambrosial and stench. And as we've seen, nearly two centuries later Crocker and Henderson—the ten-thousand-odors duo—thought there might be four dimensions: fragrant, acid, burnt, and goaty.

The list goes on and on, with many classification systems appearing bizarre when viewed from the outside. The Suya of Brazil regard odors as bland, strong, or pungent. Sounds sensible—but oddly, adult men smell bland, while women smell strong and the elderly smell pungent. The Serer-Ndut of Senegal have five categories: urinous, rotten, milky/fishy, acrid, and fragrant. Monkeys, cats, and Europeans have a urinous odor. Rotten-smelling things include cadavers (obviously), mushrooms (understandably), and ducks (um . . .); acidic smells include those of tomatoes and spiritual beings. (Prizes for anyone who can explain what tomatoes and ghosts have in common.) The Serer-Ndut themselves have a fragrant odor, the most positive of the five categories—but then again, so do onions.

Any classification system that uses words (and their underlying concepts) is bound to suffer from cultural blinkers. You name what's important to you—and that, overwhelmingly, is what's in front of your nose from day to day. "We" always smell good, and "they" smell bad. You can't understand goaty odors if you've never encountered a goat. As we'll see, professional flavorists sort odors into categories like fruity, floral, and spicy: basically, the kinds of ingredients they deal with in their daily work.

Is there any way out of this cultural trap, any way to sort odors into dimensions without having to resort to language? Andreas Keller of Rockefeller University thinks so. A big bear of a man with a soft German accent, Keller works the boundary between sensory science and philosophy, making significant contributions to each. To test the dimensionality of smell from first principles, Keller sets out three vials with different odor molecules and asks people to group them by similarity. If everyone picks the same pair of odorants as most similar, he knows that those two sit together along some dimension—they're both fruity, say. If no pair is put together more commonly than the others, on the other hand, then all three odorants must be equidistant from one another, like the points of an equilateral triangle. That means there must be at least two dimensions. Four equidistant odors require three dimensions, and so on. The concept is straightforward, even though the math gets more than a little bit hairy as the number of dimensions grows.

Keller's hope is that sooner or later, adding more odors no longer requires new dimensions. The big question is whether that happens after just a few dimensions—in which case odors really do fall into meaningful categories—or many. The worst-case scenario would be that there's a separate dimension for each of our four hundred or so odor receptors, which would effectively mean that there is no underlying structure, no effective way to group smells into perceptual categories. "I think everything up to about twenty or thirty dimensions would be interesting," says Keller. His experiments are still ongoing as I write, but he's less and less optimistic that he'll end up with a manageable number of dimensions.

Our poor performance at naming and sorting smells is, no doubt, part of the reason why most people think humans are olfactory incompetents, with noses that are good for little more than keeping our glasses from falling off. But in fact, we're too hard on ourselves. Our noses are a much more powerful tool than most of us realize—more sensitive, in many cases, than the most expensive piece of laboratory equipment.

Case in point: If you had happened to cross the University of California at Berkeley campus in the early 2000s, you might have noticed an undergraduate—blindfolded, earplugged, and wearing coveralls, knee pads, and heavy gloves—crawling across the lawn with nose to ground, zigzagging slightly back and forth. Was he rolling a peanut across the campus with his nose as punishment for some arbitrary offense during a fraternity initiation? Was he groveling before more senior fraternity brothers? No. He was following a scent trail laid down by a chocolate-soaked string—and doing it almost perfectly.

This rather odd spectacle was another of Noam Sobel's slightly skewed brainchildren. (At the time, Sobel was a junior professor at Berkeley, though he's now at Israel's Weizmann Institute of Science.) For the chocolate-tracking experiment, Sobel and his students tested a total of thirty-two people and found that twenty-one of them could find and follow the chocolate track by nose alone, with all their other senses blocked. Better yet, when Sobel gave four of the volunteers a chance to practice repeatedly, every one of them got better at following the trail, moving faster and casting about less. When the trackers tried again while wearing a nose clip, every one of them failed to find the trail—clear proof that they weren't navigating by some other cue that the experimenters had overlooked.

And it's not just that we're less worse than we thought: our noses actually compare favorably with those of other animals—even ones renowned for their sense of smell. Matthias Laska, a psychologist at Linköping University in Sweden, has been measuring the acuity of animals' noses for decades, since long before Sobel's chocolate study. The gold standard for this sort of thing is to measure the olfactory threshold, the lowest concentration of an odorant that can be detected—exactly what Doty's machine tried to measure for my nose. Since you can't just ask a monkey or an elephant whether it can smell something, Laska does the next best thing: He teaches the animal to associate the odor with a food reward—a yummy carrot for the elephant, for example, or a peanut for a squirrel monkey. Then he lets the animal choose one of two boxes: one unscented and empty, and the other bearing the telltale scent and containing the treat. If the animal picks the treat consistently, it must be able to smell the signal, and Laska repeats the test with a lower concentration of the odorant. When he gets to the point where the animal can't tell which box has the treat, he knows the odorant signal has fallen below the olfactory threshold.

Over the years, Laska has used this method on everything from bats to mice to elephants to several species of monkey. Out of curiosity, he compared his results with what other researchers had reported for humans—and noticed that the animals weren't necessarily any better smellers than we are. Intrigued, he started searching the literature for every study he could find that reported an olfactory threshold for a nonhuman animal, then looked to see if he could find a comparable threshold for humans.

The results showed that his initial comparisons weren't a fluke. Human noses are more sensitive than those of rats for thirty-one of the forty-one chemicals that have been tested on both species, for example. Humans even outperform dogs in detecting five of fifteen scents. "The traditional textbook view that humans have a poorly developed sense of smell is not warranted," says Laska. "We are not that hopeless."

If so, why do customs agents use beagles instead of Bostonians to detect drug smugglers? Why don't we track our dogs through the park as readily as they track us? Part of the difference may be that most of the time, we're distracted by our senses of sight and sound. "Except for smell researchers such as me," says Laska, "we are not constantly aware of the odor stimuli in our environment." For one thing, it's simply harder to pay attention to smells than to sights or sounds. If you're looking for a friend's face in a crowd, or scanning a bookshelf for a particular title, your vision is focused on a specific point in space. Similarly, when you're trying to listen to one conversation amid a noisy cocktail party, you'll turn to face the speaker and concentrate on that one spot. This tight spatial focus helps us notice what we see and hear.

By contrast, we don't ordinarily focus our smelling in the same way. Sure, you can stick your nose into a wine glass, or take a sniff at the back of a toddler's diaper—both instances where we really do pay attention to odors. But that's not the way we usually use our noses. For most of the day, our noses aren't focused on any one particular thing. Instead, we smell an undifferentiated mix of everything that's going on around us, the olfactory equivalent of peripheral vision with nothing in the center of focus. Even when we're trying to pay attention to a particular odor—what's that herb in this sauce?—studies show we don't get any better at detecting that target.

Subconsciously, we probably make a lot more use of smell than we think. For example, did you know that you tend to smell your hand shortly after shaking hands with someone? Well, you do. We all do. Sobel—him again—secretly filmed unsuspecting students who thought they were waiting idly to participate in a psychology experiment. The experimenter came in, introduced him- or herself—sometimes with a handshake, sometimes not—then left the room again. Within seconds, the students who had shaken hands would lift their hand to their nose and sniff it—especially if the experimenter was of the same sex as the student. "We would see people sniffing themselves just like rats," Sobel told a reporter. Clearly, we're taking in information of some sort, even though we're not aware of it. (Knowing this may forever taint your experience of greeting people.)

Sight and sound also come to us in a continuous stream, while smell comes in discrete sniffs, separated by several seconds of "olfactory silence." That may not seem like an important difference, but it is. Continuity makes it much easier to notice changing sights and sounds—and when there's a break in continuity, we often become "change blind." In one famous experiment, an actor carrying a map approached an unsuspecting pedestrian and asked for directions. Before the person finished giving the directions, an annoyingly oblivious pair of "workers"—actually accomplices in cahoots with the experimenters—barged between the two carrying a large door. While the view was blocked, a second actor took the first one's place. After the workers left, half the people simply resumed giving directions and never even noticed that they were now talking to a different person. They were blind to the change that happened during the visual gap.

If change blindness affects even frontline senses like vision, it's likely to be even more significant to our sense of smell, where the equivalent of a large door passes through after every breath. This change blindness makes the changing smellscape much harder to keep track of, and is another reason why we don't notice smells the way we do sights and sounds, says Sobel.

But there's an even simpler reason why we humans don't often pay as much attention to smells as our dogs do. Dogs' noses are down there near the ground where most of the smells are, while ours are way up in the air. Except in unusual circumstances, like Sobel's human chocolate hounds, we simply aren't aware of the rich olfactory world of scent trails that we could be monitoring.

Our noses may be poorly positioned for following scent trails on the ground, but there's another class of odors that they're perfectly positioned to appreciate: those that contribute to the flavors of food and drink. In fact, we humans might be the virtuosos of the flavor world. To understand why, we need to recognize that what we think of as "the sense of smell" is really two different senses that share the same equipment, like taxi drivers sharing a car on alternate shifts.

Until now, we've been talking mostly about smell as a process of sniffing air in through the nostrils to the olfactory epithelium. This kind of smelling tells you about what's out there in the world: fragrant flowers, burning leaves, your nearby lover. Experts call it *orthonasal olfaction*, but it's fine to think of it as just sniffing.

But there's another route that odor molecules can take to get to the olfactory epithelium: through the backdoor. This *retronasal olfaction* only happens when we eat or drink something. As we exhale, some of the odor compounds from the food or drink rise up the back of the throat and into the nasal cavity from the rear. In fact, the shape of our throat helps push food odors into our nasal cavity as we exhale. To show this, Gordon Shepherd and his colleagues used CAT scans to determine the precise shape of the nose, mouth, and throat of a fifty-eight-year-old volunteer, then used a 3-D printer to build a full-scale model of her anatomy. When they measured airflow through this model, they found that air inhaled through the nose forms an air curtain in the throat that effectively walls off the mouth, so that food particles and odor molecules from a mouthful of food aren't swept into the lungs. (That's a good, practical reason for chewing with your mouth closed: airflow through an open mouth disrupts the air curtain.) The curtain also ensures that our orthonasal sniffing isn't contaminated with food aromas from the mouth. But when we exhale, this air curtain shuts off, so that odor molecules from the mouth can eddy up into the nasal cavity and reach the olfactory epithelium. Retronasal olfaction, in other words, is all about flavor.

And according to Shepherd, retronasal olfaction is a skill that we humans are uniquely good at. Think about how the shape of a dog's head compares with yours. The dog has a long snout and its head projects forward from the neck, so that its nasal cavity sits well forward of the back of the mouth. As a result, the retronasal path to the olfactory epithelium is a long journey down a narrow tube, and relatively few odor molecules are likely to make the trip. Dogs' noses, in other words, are optimized for orthonasal smelling. In contrast, humans have relatively short noses. More importantly, our upright posture means that instead of projecting forward, our heads sit immediately above our necks, so that retronasal odor molecules just have to waft up a short way from the back of the mouth to the olfactory epithelium. It's a much shorter, easier path, and it's reasonable to think that our retronasal olfaction—and therefore, our appreciation of flavor—is correspondingly better. (We also have bigger brains to think about the flavors we taste, which further sharpens our appreciation. More on that in a later chapter.) The result is that when you sit back and appreciate the complex flavors of a soup or a glass of wine, you're doing something that few other species—perhaps none—would be capable of. We should feel special!

The existence of these two ways of smelling might explain one of the peculiarities in our experience of flavor. Most of the time, sniffing a food tells us pretty much what flavor we're going to get when we eat it—but not always. We can all think of foods—really stinky cheeses such as Limburger come to mind, and the notorious Asian fruit known as durian—that smell vile as you're trying to work up the courage to eat them, yet "taste" divine once you actually put them in your mouth. Similarly, almost everyone loves the smell of freshly brewed coffee, but not everyone likes its flavor. Those differences—one professional flavorist told

Mainland that they happen for about 15 percent of odors—would make sense if we respond differently to orthonasal smells than to retronasal ones.

Confirming this scientifically is easier said than done, because it's not easy to study retronasal olfaction. You can't just squirt a dose of coffee in someone's mouth, since that also delivers taste and touch signals that aren't there when you wave a cup of the stuff under their nose. Instead, scientists have to go all techno and thread two plastic tubes into the nose so that one opens just inside the nostrils and the other at the top of the throat. Then they can use a computer to deliver precise doses of odorant to either the orthonasal or the retronasal tube, while flowing unscented air through the other tube to avoid any telltale puff-touch signals.

These studies show that retronasal odors are indeed handled differently from orthonasal ones. For one thing, olfactory thresholds tend to be lower for smells that arrive orthonasally. That makes sense: Orthonasal delivers early warnings of changes in the environment, which would need the most sensitive detector available; retronasal, on the other hand, perceives the flavor of foods that are already in the mouth—there's plenty of stimulus there, and it only needs to pick out the distinctive features so you can identify what you're eating. And in keeping with that division of labor, retronasal odors turn out to be more effective at stimulating the brain regions responsible for processing flavor.

There's likely to be a physical reason, too, why the same food might yield a different experience orthonasally and retronasally, and that has to do with the direction of airflow. Researchers haven't worked out the details yet, but it's becoming clear that our four hundred or so odor receptors aren't scattered randomly across the olfactory epithelium, but instead are sorted into several zones with a different mix of receptors in each. In particular, our most ancient odor receptors—inherited from our fish ancestors, and tuned to water-soluble odorants, the only kind that fish could experience—are clustered right at the front of the olfactory epithelium. That means they get first crack at orthonasal odors, but are last in line for retronasal ones. By the time retronasal airflow reaches these fishlike receptors, many of the water-soluble odorants may have already dropped out, mired fruitlessly in the watery nasal passages farther back. Sure enough, Sobel (again!) has found evidence that the nose is actually sorting odors from front to back in the nose. The world smells different to each nostril, he finds, with the higher-airflow nostril more attuned to non-water-soluble odors, which the orthonasal air current carries farther back to the relevant receptors. For the same reason, odorants should sort differently when inhaling orthonasal smells than when exhaling retronasal ones. The real clincher would be if someone could show that the wonderful aromas of fresh coffee, and the obnoxious aromas of ripe cheeses and durian, tended to be water soluble so that they're more accessible orthonasally than retronasally. Unfortunately, no one has done that yet, as far as I know.

The same day that I'd talked with Shepherd in Florida about retronasal olfaction, I unexpectedly put my newfound knowledge to use. I ate dinner that night in a cheap-but-excellent Mexican restaurant not too far from my cheap-but-adequate motel. I ordered a Negra Modelo, my favorite Mexican beer, and the waiter set the bottle on the table. I was about to ask for a glass—I've always been a bit of a snob about drinking my beer out of a glass "to appreciate the flavor better"—when I recalled something Shepherd had told me that afternoon. We forget what we know about retronasal smell, he said, as soon as we sit down to eat. "Think about it. Most of the flavor is when you're breathing out." *Aha*, I thought. *The glass won't do anything for the* flavor *of the beer—it will only enrich my orthonasal experience, which is different.* I drank my beer from the bottle—and sure enough, the flavor was all there.

But what flavor was it? Let's pause that scene—me with beer bottle to mouth, enjoying the chocolate and caramel flavors of the Negra Modelo—and ask whether someone else would have the same flavor experience. We already know that people differ in their taste receptors, so that your experience of the beer's hoppy bitterness could be different from mine. And we already know that even people who taste a lot of bitter—like me—sometimes learn to love it in their beer. But since the lion's share of flavor comes through retronasal olfaction (remember the jelly bean test!), it's also worth looking at how people differ in their sense of smell.

We've already seen that people have about four hundred odor receptors, more or less. Here's where things get interesting. Of those four hundred, about half work in everyone, so all of us can smell the molecules they target. The other half work in some people and not others, which means there's a huge range of stuff that some of us can smell and others can't. To further complicate things, even the working receptors often have small genetic differences from person to person, so that you might be more sensitive to certain odors than I am, and vice versa. In fact, the sample of one thousand genomes showed that you and I are likely to have meaningful differences—that is, differences that affect odor detection—in about 30 percent of our odor receptors. That means your flavor world is different from mine, and from your best friend's, and even from your parents'. Chances are that no two people (except, perhaps, identical twins) share exactly the same sense of smell. Every one of us lives in their own unique flavor world.

Not only does each person have their own distinctive set of working and broken odor receptors, but every person's nose probably mixes its receptors in different proportions. The evidence for this comes from Darren Logan, a virtuoso molecular geneticist at the Sanger Institute in Cambridge, England. Logan is a slender, compact bundle of energy with trendy glasses, dark hair cut in a short buzz, and a fascination with olfactory receptors. In particular, he's used gene-sequencing technologies to measure the abundance of each of the hundreds of olfactory receptors in the nose. There's a catch, though: To properly census an individual's complete repertoire of receptors, he needs to study entire noses—or, more precisely, entire olfactory epithelia. It's hard to convince a living person to sacrifice their sense of smell for science, and tissue from cadavers, even fresh ones, hasn't been good enough. So Logan works on mice instead.

Mice use all 1,099 of their working odor receptors in their nose—but not in equal proportion, Logan finds. Instead, a few of the receptors are very common, a slightly larger number are moderately common, and most are rare. And that pattern seems to be dictated by genes. One of the advantages of working with mice is that you can pull out a catalog from a mouse-supply company and buy as many genetically identical animals as you want, from any of several very different strains. Sure enough, when Logan compares two genetically identical mice, they have exactly the same pattern of odor receptor frequencies. In other words, when it comes to the mix of odor receptors in a nose, genes rule. Pick a different mouse strain, and the pattern is much different. Take a mouse from a different subspecies, and the differences are bigger still, with half the receptors differing in abundance by as much as a hundredfold. "That means one strain is, in theory, a hundredfold more sensitive to whatever that receptor is detecting," says Logan.

We have to be cautious about extrapolating from mice to people—plenty of researchers have ended up with egg on their face from doing that too glibly—but if people are like mice in this respect, then not only do you and I have slightly different sets of working odor receptors but we're probably genetically programmed to mix them in different proportions. If so, then the olfactory chord that coffee sounds in your brain might be richer in the horns, while mine is richer in the strings. That would help to make your flavor world even more different from mine. As I write this, Logan is trying hard to secure fresher, better human olfactory epithelia so that he can test this idea directly. He's got nine so far, donated by living people who were about to lose them anyway as a result of treatment for a rare cancer, but he needs a lot more. Stay tuned.

All this talk of genetics makes it easy to assume that where the sense of smell is concerned, you're stuck with the hand nature dealt you. To some extent that's true, of course; if you only have broken copies of a particular odor receptor gene, you're never going to be able to make use of that receptor. But the reality is a bit more complex than that. Just ask Charles Wysocki.

Wysocki has been at Monell since the 1970s, making him one of the center's longest-serving researchers, and right from the beginning he's been fascinated by individual differences in the sense of smell. (For what it's worth, early in his career, he also published a paper on how to tell male newborn mice from females.) It was Wysocki, together with Gary Beauchamp, who showed more than thirty years ago that a person's genes help determine whether they can smell androstenone, a musky, urinous-smelling compound that male boars use to signal their virility and is also a key flavor component in truffles. Their study was one of the first clear proofs that genes affect the sense of smell. But along the way, they learned something else, too.

Now semiretired, Wysocki is a small man with a slight stoop, thick gray hair, and an extravagant organ-grinder's mustache. "I started working with the compound in 1978," he recalls. "I could not smell it at all—was totally oblivious to it. I just had to trust the scales, the balances, that I was making the right stuff." After a few months of working with the compound daily, he started noticing a new odor around the lab. To his surprise, the culprit turned out to be androstenone. Somehow, he had acquired the ability to smell it. And he wasn't the only one—some of his technicians reported the same thing. Intrigued, he tested a larger sample of people. Sure enough, half of the nonsmellers became much more sensitive after a few weeks' exposure to the compound. "These people went from a nonsmeller to a pretty sensitive smeller," he says—though they never got to be as good as the best natural smellers, who can detect androstenone at just a few parts per trillion.

The picture gets even more complicated. Wysocki has tried the same experiment with other odorants, such as the sweaty-smelling 3-methyl-2-hexenoic acid, and found no change in detection ability. His colleague, Pam Dalton, showed that repeated exposures to Maraschino-cherry-smelling benzaldehyde do lead to improvement—but only in women of reproductive age, not in men, young girls, or postmenopausal women. Even now, nearly three decades later, Wysocki's not sure why some people get better at detecting some odors after being exposed to them, while other people do not.

Some of the answer, no doubt, has to do with the odor receptors themselves, and the way they interact with the odorants. But some, too, must depend on the way your brain processes odor information. People who measure olfactory thresholds quickly learn that they don't stay put. Your detection threshold for a particular odor can vary many thousandfold from one test to the next—and it doesn't matter whether the tests are separated by thirty minutes or more than a year. That's probably at least partly because our noses don't always claim the same share of our conscious attention.

That's not to say we can't get better at recognizing and identifying odors. Practice clearly helps. You can prove this yourself by pulling a bunch of spices off your kitchen shelves and trying to identify them with your eyes closed. After a few rounds of trial and error, you'll get better. The most vivid demonstration of the power of practice, though, comes from wine experts, who are much better than the rest of us at putting names to the aromas rising from their tasting glasses. But whenever scientists have tested such experts (which doesn't happen often—what wine pundit would run the risk that their nose might be shown up as below average?) they've found that their olfactory ability is nothing special. What gives their ordinary noses the ability to perform extraordinary feats of perception is simply experience. That's encouraging news for anyone who wants to sharpen their flavor senses.

If you've booked a table at the best restaurant in town, or plan to open a treasured bottle of wine, there may be other ways to amp up your sense of smell to wring more flavor out of the experience, though you may look a bit odd in the process. Nasal sprays containing sodium citrate or a compound called EDTA bind calcium ions in the mucus layer coating the olfactory epithelium, and this makes olfactory cells more sensitive for a few minutes before things return to normal.

If the thought of spritzing your schnozz every fifteen minutes at the French Laundry restaurant in California is a bit off-putting, here's another option: Use one of those nasal dilator strips that professional athletes wear across the bridge of the nose to hold their nostrils open. Athletes use them to help inhale more air faster, but as a side effect the dilators also improve airflow to the olfactory epithelium. Tests have shown that this makes it easier to detect and recognize odors.

But even if experience can alter our odor perceptions somewhat, it's clear that our genetic makeup of odor receptors is driving the flavor-perception bus. It's not just a question of odor receptors, though. More than a thousand other genes affect what happens in the sensory pathway after an odor binds to its receptor. Differences in these genes undoubtedly mean that some people have a more acute sense of smell overall than others do, just as some people see or hear better than others. Not many researchers have measured how big these differences in general olfactory sensitivity are, alas, so the subject remains a big question mark.

We're just starting to understand how these differences in genetic makeup affect our experience of flavor. For example, many people—but not all—notice a distinctive asparagus-type odor to their urine shortly after eating that vegetable. Proust noted that asparagus "transforms my chamber pot into a flask of perfume." For many years, scientists assumed that the smelly pee folks digested asparagus in a way that produced an odorous molecule called methanethiol, while the others—can we call them sweet pees?—did not. But in 1980, researchers fed a pound of canned asparagus to a sweet-pee volunteer, collected his urine afterward, and offered a whiff to unsuspecting volunteers. To the researchers' surprise, they found that anyone who could smell asparagus in their own urine could also detect it in the supposedly sweet pee of the donor. In other words, the difference between sweet and smelly is not in the digestion of the eater, but in the nose of the smeller. We now know that a particular odor receptor, *OR2M7*, may be responsible. (Actually, it turns out that there are a few people who really do produce odorless urine, for unknown reasons.)

It's likely that differences in odor perception help explain why people like different foods. Take cilantro, for example. Most people love its bold, grassy flavor, but a substantial minority detests the stuff, describing its flavor as soapy or "buglike." (How do they know that, I wonder?) Scientists at the personal-genomics company 23andMe recently linked this preference to differences in or near the *OR6A2* gene.

But on closer examination, there's a cautionary story here for anyone who'd like to believe that genes are destiny. If every person with one variant of *OR6A2*—let's call it variant X—loved cilantro and every person with variant Y hated it, then we'd say that *OR6A2* explained 100 percent of the difference in perception. If *OR6A2* had no effect at all, of course, it would explain 0 percent of the difference. The closer to 100 percent you get, the stronger the effect. For cilantro preference, *OR6A2* turned out to explain less than 9 percent of the difference in perception. In practical terms, *OR6A2* isn't telling us much about cilantro preference at all.

A lot of olfactory genetics turns out to be like that, as I learned when I asked Joel Mainland to profile my own olfactory genes. Since scientists know only a handful of olfactory receptor genes so far that affect perception, this involved identifying my variants of just a few OR genes, rather than doing my whole genome sequence. A few weeks later, I visited his lab and sat through a battery of olfactory tests where I rated the intensity and pleasantness of the odors targeted by those genes.

The results could only be described as disappointing. Take *OR11A1*, for example. This OR detects an earthy-smelling molecule called 2-ethylfenchol that sometimes appears as an off flavor in beer and soft drinks. Three variants, or alleles, of this OR are common in the human population, one that is good at detecting 2-ethylfenchol and two that are relatively poor at the job. Mainland's peek at my genome showed that I have two copies of the sensitive allele, which should make me an especially sensitive smeller of 2-EF. And, since people tend to find stronger odors less pleasant, Mainland predicted that I'd rate 2-EF more unpleasant than the average person.

In fact, though, both predictions turned out wrong. On a scale from 0 (undetectable) to 7 (overpoweringly intense), I rated the intensity of 2-EF as 3.4, much less than the 4.8 that Mainland predicted. I gave it a 5.0 for pleasantness (evidently I like the smell of dirt), while Mainland predicted I'd rate it just 3.2, or mildly unpleasant. Other pairs of receptor and odor, such as *OR10G4* and smoky-smelling guaiacol, *OR11H7* and cheesy/sweaty-smelling isovaleric acid, and *OR5A1* and floral-smelling beta-ionone gave similarly confusing results. Occasionally things worked out more clearly. I've got one functional and one broken copy of *OR7D4*, the receptor that detects androstenone, the boar-urine-and-truffle odor that Wysocki studied. That should leave me as a moderate smeller and an enthusiast for truffles—which, in fact, I am. But it doesn't always turn out that way, says Mainland. "We have a lot of people who have two functional copies and don't smell it, and we have some people who have two nonfunctional copies and still smell it."

It's not surprising that single genes do such a poor job of predicting flavor perceptions, says Mainland. Since most odorants trigger more than one receptor, my response to any given odor probably depends on my genetic makeup at several genes. That muddies the waters a lot. "I'm dividing you based on one receptor, but you also have 399 other receptors, and that's a lot of noise to push through," he says. For example, he and his colleagues have found that *OR10G4* can detect both guaiacol and vanillin, the key molecule in the aroma of vanilla, but it's much more sensitive at detecting the former. When they look more closely, they find that people with a damaged copy of *OR10G4* tend to report that guaiacol smells less intense, but report no difference in vanillin—which, presumably, depends largely on another receptor instead. Clearly, linking genetics to flavor perception still has a long, long way to go.

What we'd really like, of course, is to understand olfaction well enough so we can reproduce smell sensations artificially, as we do for sights and sounds. When Luke Skywalker's X-wing Starfighter destroys Darth Vader's Death Star, we see Luke in the cockpit, even though what's really before us is just pixels on a screen. That's because we know how to make a video image that mixes those pixels in a way that our eyes and brain interpret just the same as if it were really happening. We hear the explosion, even though none really occurred (and no sound waves would travel through the vacuum of space, but that's a different issue), because we know how to re-create sounds from a string of zeros and ones in a digital file.

We're nowhere near being able to do that for flavor. Sure, you can find a few oddball episodes in cinematic history where people have re-created—or, more precisely, imported—specific odors that fit the scenes of a movie. Take Smell-O-Vision, for example. In 1960, film producer Mike Todd Jr. (Elizabeth Taylor's stepson) employed a system for mechanically releasing odors into a movie theater during the movie *Scent of Mystery*. Audiences were supposed to get a whiff of pipe smoke, for example, when one particular character appeared onscreen. The system cost tens of thousands of dollars per theater—a lot of money in 1960—and didn't work very well. In 2000, readers of *Time* magazine voted Smell-O-Vision one of the "Top 100 Worst Ideas of All Time." Still, that hasn't kept novelty-seeking filmmakers from trying again now and then, albeit usually using scratch-and-sniff cards instead of a forced-air system.

But all of these novelties merely used odors prepared ahead of time. In that sense, they're the equivalent of showing someone a photocopied picture. The real goal of digital olfaction—being able to make up any smell (and hence, any flavor) you want, to order, by combining elements from a small set of "primary odors"—was nowhere in sight.

Today, several decades later, that goal might just be visible. At the very least, we can estimate the scale of the problem. Every smell on Earth must be encoded by some combination of our four hundred-odd odor receptors. In theory, then, an arsenal of about four hundred primary odorants, each of which tickled a different odor receptor, should allow you to mix and match to re-create any smell. In practice, the task ought to be somewhat simpler than that, because it's likely that some of our odor receptors are redundant copies. For anyone interested in digitizing only flavor-related odors, the field gets a little narrower yet, since we can ignore all the receptors that are never activated by food odorants. In fact, Mainland thinks it should be possible to get at least a rough sketch of the odors—and therefore the flavors—of the vast majority of foods with many fewer primaries than that. He's been working with a flavorist at Coca-Cola who claims that with just forty primaries, you can get a recognizable facsimile of 85 percent of all foods.

When I visited Mainland's lab, he screwed the top off a vial and gave me a whiff of the concoction. "Do you recognize this?" he asked. It certainly smelled familiar, but—as so often happens when we try to identify smells cold, without any prompting—I found myself tongue-tied and unable to put a name to it. Once he told me—strawberry—it all snapped into place: Of course, strawberry! It was indeed a recognizable, though not perfect, imitation. A real strawberry contains hundreds of scented molecules. But with just four of these—cis-3-hexenol (cut grass), gamma-decalactone (waxy), ethyl butyrate (generic fruitiness), and furaneol (caramelized sugar)—Mainland can build a mix that smells recognizably like strawberry. It's not perfect, he admits—more like a pixelated image than a high-resolution version. However, he says, "We're okay with eight-bit graphics that gives you a sketch of what's going on. If we're making a poor strawberry, but it's still strawberry and not cherry or banana, we're happy with that."

Even if he could match the real thing perfectly, people might not realize it. "The problem we have is that everybody tells us it's a terrible strawberry," Mainland says of his facsimile. "But if you smash up a strawberry and put it in an olfactometer, people will also tell us it's a terrible strawberry." In our day-to-day lives, it turns out, we don't generally notice all the components of a familiar odor, so we often don't have a very good mental image of what the real thing smells like—especially when we lack the visual context. People don't usually notice the green, vegetative note in strawberry, for example, so its presence in a crushed-up real strawberry can strike them as false, somehow.

So far, all of Mainland's efforts to mimic strawberry or blueberry or orange aromas use odor components that are naturally found in the target aroma. Ideally, he'd like to go one better someday. "What we would really like to do is make strawberry without anything that's in a real strawberry," he says. To that end, he's intrigued by a molecule known to chemists as ethyl methylphenylglycidate, a mouthful unpronounceable without sounding like Sylvester the cat spluttering at Tweety Bird. To flavorists, the chemical is known as "strawberry aldehyde." As you might guess from the name, it has a strawberrylike aroma and is often used as an artificial strawberry flavor, even though it doesn't occur in a real strawberry. (You can't always trust a name, though—despite its moniker, strawberry aldehyde isn't actually an aldehyde.) Mainland would love to know whether strawberry aldehyde activates the same odor receptors as the components of real strawberry odor, to see whether that accounts for its mimicry.

But what if you wanted not just eight-bit graphics but a high-resolution image that reproduces the real flavor precisely? So far, the closest approach to this ultimate goal comes from a recent German study led by Thomas Hofmann at the Technical University of Munich. In what can only be described as a heroic assault on the university library, Hofmann and his colleagues (including the delightfully named Dietmar Krautwurst, who was clearly destined for a career in food science) read through more than sixty-five hundred scientific books and papers that analyzed the flavor molecules present in particular foods. They winnowed these down, selecting only the best and most detailed studies, until they ended up with more than two hundred food items—everything from mushrooms to taco shells, Scotch whisky to donuts—for which the key odorant molecules had been identified. Most of the papers even took things one step further by showing that a mix of those key odorants smelled indistinguishable from the real item.

The surprising thing is that the aromas, and hence the flavors, of all these diverse foods could be re-created using a palette of just 226 key odorants. That's remarkably encouraging, given the thousands of smelly molecules present in that range of foods. Some of these key odorants are what they call "generalists" that turn up over and over again. The cooked-potato-smelling methional, for example, figures in the odor of more than half the foods, while green-grassy hexanal and fruity-fresh acetaldehyde play a role in 40 percent and 29 percent, respectively. Many other odorants contributed a distinctive note to just a few food items, such as garlic's diallyl disulfide and grapefruit's 1-p-menthene-8-thiol.

Sometimes, they found, it takes only a handful of key odorants to replicate a food's flavor. Cultured butter, for example, needs only three: the buttery-smelling generalist butan-2,3-dione, coconut-like delta-decalactone, and sweaty-smelling butanoic acid. Other foods, like beer and cognac, required eighteen and thirty-six key odorants, respectively, to mimic their bouquet precisely—a lot, but still just 10–15 percent of the total set of primaries.

Of course, trying to build a digital-flavor unit with 226 primaries is still a huge technical challenge. But if you could do it—even if it took trained technicians and an expensive, well-stocked lab—then the sense of smell (and, by extension, much of flavor itself) would finally free itself from subjectivity and be on a truly objective footing. We could take an olfactory "snapshot" of a ripe Georgia peach or a tomato fresh from the garden in the heat of August and reproduce it exactly. We could save a famous chef's signature dish and archive it in a museum. And we could collect flavor memories of our travels and revisit them at home, just as we now do with photographs.

There's a lot of work still to be done before those fantasies can become realities. And not just on the olfactory front, either. As it turns out, there's more to flavor than just taste and retronasal smell. The physical sensations of touch—texture, temperature, and the like—also play a huge role in it.

Chapter 3

THE PURSUIT OF PAIN

I've been procrastinating. On my dining room table I have lined up three hot peppers: one habanero, flame-orange and lantern-shaped; one skinny little Thai bird's eye chili; and one relatively innocuous jalapeño, looking by comparison like a big green zeppelin. My mission, should I choose to accept, is to eat them. For you, dear readers.

In ordinary life, I'm at least moderately fond of hot peppers. My fridge has three kinds of salsa, a bottle of Sriracha, and a jar of Szechuan hot bean paste, all of which I use regularly. But I'm not extreme: I pick the whole peppers out of my Thai curries and set them aside uneaten. And I'm a habanero virgin. Its reputation as the hottest pepper you can easily find in the grocery store has me a bit spooked, so I've never cooked with one, let alone eaten it neat. (In fact, the first habanero I bought went moldy in my fridge while I was working up my courage.) Still, if I'm going to write a chapter that's largely about hot peppers, I ought to have firsthand experience at the high end of the range. Plus, I'm curious, in a vaguely spectator-at-my-own-car-crash way.

When people talk about flavor, they usually focus on taste and smell, just as we have so far. But there's a third major flavor sense, as well, one that's often overlooked: the physical sensations of touch, temperature, and pain. The burn of chili peppers is the most familiar example here, but there are others. Wine mavens speak of a wine's "mouthfeel," a concept that includes the puckery astringency of tannins—something tea drinkers also notice—and the fullness of texture that gives body to a wine. Gum chewers and peppermint fans recognize the feeling of minty coolness they get from their confections. And everyone knows the fizzy bite of carbonated drinks.

None of these sensations is a matter of smell or taste (though of course Coca-Cola also tastes sweet and smells of caramel, citrus, and other flavors). In fact, our third primary flavor sense flies so far under our radar that even flavor wonks haven't agreed on a single name for it. Sensory scientists are apt to refer to it as "chemesthesis," "somatosensation," or "trigeminal sense," each of which covers a slightly different subset of the sense, and none of which mean much at all to the rest of the world. The common theme, though, is that all of these sensations are really manifestations of our sense of touch, and they're surprisingly vital to our experience of flavor. Taste, smell, touch—the flavor trinity.

Sensory scientists have known for decades that chili burn is something different from taste and smell—something more like pain. But the real breakthrough in understanding chili burn came in 1997, when pharmacologist David Julius and his colleagues at the University of California, San Francisco, finally identified the receptor for capsaicin, the active ingredient in chili heat. The task demanded a lot of patience: Julius and his team took every gene active in sensory nerve cells, which respond to capsaicin, and swapped them into cultured kidney cells, which don't. Eventually, they found a gene capable of making the kidney cells respond. The gene turned out to encode a receptor—eventually named *TRPV1*, and pronounced "trip-vee-one"—that is activated not just by capsaicin but also by dangerously hot temperatures. In other words, when you call a chili pepper "hot," that's not just an analogy—as far as your brain can tell, your mouth really is being burned. That's a feel, not a smell or taste, and it passes to the brain through nerves that handle the sense of touch. Like other touch receptors, *TRPV1* receptors are found all over the inner layer of your skin, where they warn you of burn risk from midsummer asphalt, baking dishes straight from the oven, and the like. But they can only pick up pepper burn where the protective outer skin is thin enough to let capsaicin enter—that is, in the mouth, eyes, and a few other places where the sun doesn't shine. This explains the old Hungarian saying that "good paprika burns twice."

Further tests showed that *TRPV1* responds not just to heat and capsaicin but to a variety of other "hot" foods, including black pepper and ginger. More recently, several more *TRP* receptors have turned up that give other food-related somatosensations. *TRPA1*, which Julius calls the "wasabi receptor," causes the sensation of heat from wasabi, horseradish, and mustards, as well as onions, garlic, and cinnamon. *TRPA1* is also responsible for the back-of-throat burn that aficionados value in their extra-virgin olive oil. A good oil delivers enough of a burn to cause a catch in your throat and often a cough. In fact, olive oil tasters rate oils as "one-cough" or "two-cough" oils, with the latter getting a higher rating. (One reason wasabi feels so different from olive oil is that the sulfur-containing chemicals in wasabi are volatile, so they deliver wasabi's characteristic "nose hit," while nonvolatile olive oil merely burns the throat. Olive oil may also trigger *TRPV1* receptors to some extent.) Curiously, *TRPA1* is also the heat receptor that rattlesnakes use to detect their prey on a dark night.

TRPM8—which is triggered by cool temperatures, rather than warm ones—is responsible for the cooling sensation of menthol (from mint) and eucalyptol (from eucalyptus). Food companies love *TRPM8*, by the way, because there's a huge market for the mouthfeel of coolness in, for example, gums and mouthwashes. In fact, the menthol that used to make chewing gum minty fresh has largely been replaced by other molecules that do a more effective job of triggering *TRPM8*, so that your gum holds its minty flavor longer.

Now that scientists understand the receptors that account for these perceptions, we can begin to unpack some of the subtleties of chili pepper flavor. Chili aficionados get pretty passionate about their pods, choosing just the right kind of chili for each application from the dozens available. The difference among chili varieties is partly a matter of smell and taste: Some are sweeter, some are fruitier, some have a dusky depth to their flavor. But there are differences in the way they feel in your mouth, too.

One difference is obvious: heat level. Chili experts measure a chili's level of burn in Scoville heat units, a scale first derived by Wilbur Scoville, a pharmacist and pharmaceutical researcher, in 1912. Working in Detroit—not exactly a hotbed of chili cookery, especially back then—Scoville had the bright idea that he could measure a pepper's hotness by diluting its extract until tasters could no longer detect the burn. The hotter the pepper was originally, the more you'd have to dilute it to wash out the burn. Pepper extract that had to be diluted just tenfold to quench the heat scores ten Scoville heat units; a much hotter one that has to be diluted one hundred thousandfold scores one hundred thousand Scovilles. Nowadays, researchers usually avoid the need for expensive panels of tasters by measuring the chili's capsaicin content directly in the lab and converting that to Scoville units. The more capsaicin, the hotter the chili.

However you measure it, chilis differ widely in their heat level. Anaheims and poblanos are quite mild, tipping the scale at about 500 and 1,000 Scovilles, respectively. Jalapeños come in around 5,000, serranos about 15,000, cayennes about 40,000, Thai bird's eye chilis near 100,000, and the habanero on my table somewhere between 100,000 and 300,000 Scovilles. From there, intrepid souls can venture into the truly hot, topping out—as of this writing—with the Carolina Reaper at a staggering 2.2 million Scovilles, which approaches the potency of police-grade pepper spray. Check out the online videos of hardy souls eating these peppers. "Successful" consumers sometimes end up in an ambulance. "They're in pain, they swallow, they retch for a couple of hours. I miss the point," says Bruce Bryant, who studies hot peppers and other mouthfeel perceptions at the Monell Chemical Senses Center. "I'm not a big pepper eater," he adds. "I was 30 years ago, but I don't have anything to prove any more about how much pain I can stand."

For those with an extreme mind-set, it's worth noting that pure, unadulterated capsaicin packs a whopping sixteen million Scovilles. If you're looking for the ultimate in hot, though, a chemical called resiniferatoxin—found in a Moroccan spurge plant—delivers sixteen *billion* Scovilles in its purified form. That's easily enough to cause lethal chemical burns. People don't eat the stuff.

Returning to the culinary realm, many chili heads claim that a pepper's heat is defined by more than just intensity. If anyone would know about this, I figured, it would probably be Paul

Bosland, the director of the Chile Pepper Institute at New Mexico State University. Bosland's institute has its fingers in all things chili—and, as a plant breeder by trade, he has a keen professional interest in all the tiny details of how chili heat differs from one pod to the next.

Bosland says he and his colleagues distinguish four other components to chili heat in addition to heat level. The first is how fast the heat starts. "Most people, when they bite the habanero, it maybe takes 20 to 30 seconds before they feel the heat, whereas an Asian chili is immediate," he says. Chilis also differ in how long the burn lasts. Some, like jalapeños and many of the Asian varieties, fade relatively quickly; others, like habaneros, may linger for hours. Where the chili hits you also varies. "Usually, with a jalapeño, it's the tip of your tongue and lips, with New Mexico pod types it's in the middle of the mouth, and with a habanero it's at the back," says Bosland. And fourth, Bosland and his crew distinguish between "sharp" and "flat" qualities of burn. "Sharp is like pins sticking in your mouth, while flat is like a paintbrush," he says. New Mexico chilis tend to be flat while Asian ones tend to be sharp—a quality I certainly noticed the last time I ate Thai food.

Some of that difference no doubt arises because chili burn is about more than just capsaicin. In fact, there are at least twenty-two different chemicals in the capsaicin family, each of which tickles the *TRPV1* receptors a little differently. Nordihydrocapsaicin, for example, seems to burn more at the front of the mouth, while homodihydrocapsaicin hits more at the back of the throat, explains Michael Mazourek, a chili breeder at Cornell University. But the difference matters less than you'd think. Nordihydrocapsaicin, for example, is only half as potent as capsaicin, and makes up at most 7 percent of the capsaicin content. "So you can see its vote doesn't count for much, even though it would have some different properties," says Mazourek. The two most common capsaicinoids—capsaicin itself and dihydrocapsaicin, which together make up 50–90 percent of the capsaicinoids in every pepper variety—are also by far the most potent, so they pretty much run the show.

A bigger reason why some peppers feel different from others, Mazourek thinks, may be something that food scientists call "matrix effects." In order to perceive a pepper's heat, the capsaicin has to get out of the pepper's cells and onto your tongue, lips, or palate. Pepper varieties with tougher cells would release their capsaicin more slowly as you chew, giving you a burn that builds more slowly, and perhaps hits you farther back in the mouth. Oil content might affect how quickly the capsaicin washes away, thus affecting the burn's duration.

None of this explains the difference between "sharp" and "flat" chili heat, though. In fact, no one I talked to had a good explanation for this phenomenon. Mazourek and Julius both punted. Bosland suggested that different capsaicinoids could be involved, but he admitted he had no data to back up that conjecture. And Bryant wondered whether everyone is just fooling themselves into thinking there's a difference. "All you need to do is tell somebody that it's going to be different," he says. "I'm really skeptical about some of these reports."

Okay, enough theory. I've put off my chili tasting as long as I could, and now it's time to take the plunge. First up, the jalapeño. As you'd expect from its relatively wimpy ranking in the hot pepper standings, it gives only a mild burn, which builds gently and mostly at the front of the mouth. (Score one point for Bosland.) Confronted with such a tame burn, I have plenty of attention left to focus on its thick, crisp flesh and sweet, almost bell-peppery flavor.

The Thai bird's-eye chili, second on my list, is much smaller, and its flesh proves to be much thinner and tougher. Despite that, though, it almost immediately lets loose a blast of heat that explodes to fill my mouth from front to back, making me gasp for breath. No gradual build to this one—it's a sledgehammer blow. If I think hard, I might imagine that the chili heat is a little bit sharper, pricklier, than the jalapeño. But I could just be fooling myself.

Finally, the one I've been dreading, the habanero. I cut a tiny slice (call me a coward—I never signed up for the full three-hundred-thousand-Scoville experience) and start chewing. The first thing that strikes me is how different the flavor is. Instead of a vegetal, bell pepper flavor, the habanero gives me a much sweeter, fruitier impression that's surprisingly pleasant. For about fifteen or twenty seconds, anyway—and then, slowly but inexorably, the heat builds. And builds. And builds, long after I've swallowed the slice of pepper itself, until I can't think of much else besides the fire that fills my mouth. It definitely hits farther back in the mouth than the Thai chili, though there's a late-breaking flare-up on my tongue as well. The whole experience lasts five or ten minutes, and even now—a good half hour later—it's as though coals are gently banked in my mouth. Wow.

Having set my mouth afire, I'd now like to quench the burn. Surprisingly, scientists can't offer a whole lot of help in this regard. A cold drink certainly helps, because the coolness calms the heat-sensing *TRPV1* receptors that capsaicin excites. The only problem—as you've no doubt noticed if you've tried to cope with a chili burn this way—is that the effect goes away in just a few seconds, as your mouth returns to normal body temperature.

You've probably heard, too, that sugar and fat help douse the fire, but the researchers themselves aren't entirely convinced. "The best thing out there is probably cold, whole milk," says John Hayes of the University of Pennsylvania. "The cold is going to help mask the burn, the viscosity is going to mask the burn, and the fat is going to pull the capsaicin off the receptor." When pressed, though, he notes that there's not a lot of data to back that up. Making a food more viscous has been shown to damp down taste—probably just because it provides a competing sensation to distract our attention, he notes, but he can't think of anyone who's tested whether it also reduces chili burn. And he's not entirely sure that sugar really helps, either. "I'm not convinced that it actually knocks the heat down, or whether it just makes it more pleasant," he says. Even the value of fats or oils—which sounds like they ought to help wash capsaicin, which is fat soluble, off the receptors—is in dispute. If you're feeling the burn, says Bryant, the capsaicin has already penetrated your tissue, so a superficial rinse of whole milk or olive oil isn't going to help much. Instead, Bryant has another suggestion: "Go kick a brick wall or take a hammer to your thumb. You'll forget all about your tongue," he says.

Bryant's tongue is firmly in his cheek there. At least I think so, since I don't know of any hot pepper fans who are actually tempted to smash their thumb with a hammer to cope with the burn. In a way, that highlights what might be the most fascinating aspect of chili peppers, wasabi, and their ilk. Millions of people actively seek out the pain of hot chilis as a form of pleasure. The burn features prominently in more than a few of the world's great cuisines, with more than a quarter of the world's population eating hot peppers daily. More than three quarters of Americans in one recent survey expressed an interest in eating hot chilis, and the British spend £17 million (approximately $22 million, at the time of this writing) annually on hot sauce.

And yet, only the oddest among us makes a similar fetish of inflicting other sorts of pain. We don't take pleasure in eating food that's still searingly hot from the oven, even though that delivers exactly the same sensation we get from chilis: same receptors, same nerves. We don't choose to chemically burn our tongues with strong acids. We don't hammer our thumbs for fun. So why do we happily, even eagerly, inflict pain by chilis? Whatever the secret is, it seems to be unique to humans. No other mammal on the planet has a similar taste for chilis. (Birds eat them enthusiastically, but only because they lack receptors that respond to capsaicin. To a parakeet, the hottest habanero is as bland as a bell pepper.)

One possible explanation is that chili lovers simply don't feel the pain as intensely as those who shun hot peppers. In the lab, it's certainly true that people who are repeatedly exposed to capsaicin become less sensitive to it. In fact, liniments such as Heet and Icy Hot contain capsaicin for its painkilling properties. It's easy to imagine that something similar might be happening to chili eaters, too, because people who eat hot peppers more often report less burn from a given test dose of capsaicin. But when you look more closely at that result, it gets less convincing. For one thing, an inexperienced chili eater might think a sample is the hottest thing they've ever eaten and rate it nine out of ten for burn ("I could light a cigarette from my mouth!" a shocked Romanian friend exclaimed once during an Indian dinner), while an old hand might say, "Hah, I've had hotter," and give the same sample a five. Alert researchers can avoid this problem by following Linda Bartoshuk's lead and pegging the upper end of the scale as "the most intense sensation of any kind you've ever experienced," but not every study does so.

Genetics may play some part, too. Studies of identical twins (who share all their genes) and fraternal twins (who share only half) suggest that genes account for between 18 and 58 percent of our liking for chili peppers. Some people may have more sensitive *TRPV1* receptors, for example—though Hayes, who's looking into that now, says, "The jury is really still out on whether there is meaningful *TRPV1* variation." Similarly, some (but not all) studies have found that chilis cause more pain in supertasters or those with more fungiform papillae (and therefore, presumably, more pain nerve endings) on their tongue.

It's abundantly clear, though, that chili lovers aren't immune to the pain. Just ask one. "I like it so all my pores open up and tears are rolling down my face," says Hayes. "But with two young kids in the house, I don't get that very often." For now, Hayes gets his pain mostly from a handy bottle of Sriracha hot sauce. "My kids refer to it as Daddy's ketchup," he says.

It's clear from listening to Hayes that he—and probably most other chili eaters—actively enjoys the pain. That paradox has drawn the attention of psychologists for several decades now. Back in the 1980s, pioneering chili researcher Paul Rozin of the University of Pennsylvania proposed that chili eating is a form of "benign masochism," like watching a scary movie or riding a roller coaster. After all, most forms of pain are warnings of imminent harm. That baked potato still steaming from the oven is hot enough to kill the cells lining your mouth, potentially causing permanent damage. The hammer landing on your thumb can break bones. But chili burn—except at its uppermost, million-Scoville extreme—is a false alarm: a way to get the thrill of living on the edge without the risk of exposing yourself to real danger.

A few decades later, Hayes and his student Nadia Byrnes (perhaps the best name ever for a hot pepper researcher) took Rozin's ball and ran with it. If chili heads are looking for thrills, Byrnes and Hayes reasoned, you'd expect them to have sensation-seeking personalities. And, sure enough, when they went to the vast arsenal of "instruments"—that is, personality tests—that psychologists have developed to measure facets of personality, they found several measures of sensation seeking, of which the latest and best was the Arnett Inventory of Sensation Seeking. Then they set out to see whether chili lovers really do crave excitement.

As a chili eater, I have a personal stake here, so I found the Arnett Inventory on the Internet and took the test myself. It's only twenty questions long. Each question gives a statement about yourself (examples include "When I listen to music, I like it to be loud," "It would be interesting to see a car accident happen," and "I would have enjoyed being one of the first explorers of an unknown land") and asks you to score it on a four-point scale from "does not describe me at all" to "describes me very well." Add up the scores, and there you have it: one number, somewhere between twenty and eighty, that summarizes your yen for stimulation. Of such bricks is the edifice of personality research built. (Actually, Arnett gives you two subscores, as well: one for novelty seeking, and the other for intensity seeking. I scored high—thirty out of forty—for the first, and very low—just nineteen—for the second. I'm not a psychologist, and self-diagnosis is dubious in any case, but that fits: I'm eager and willing to visit a new place or eat a new kind of food, but terrified by roller coasters and irritated by overly loud music.)

Sure enough, when Byrnes and Hayes tested nearly 250 volunteers, they found that chili lovers were indeed more likely to be sensation seekers than people who avoided chilis. And it's not just that sensation seekers approach all of life with more gusto—the effect was specific to chilis. When it came to more boring foods like cotton candy, hot dogs, or skim milk, the sensation seekers were no more likely to partake than their more timid confreres.

Chili eaters also tended to score higher on another aspect of personality called sensitivity to reward, which measures how drawn we are to praise, attention, and other external reinforcement. And when the researchers looked more closely, an interesting pattern emerged: sensation seeking was the best predictor of chili eating in women, while in men, sensitivity to reward was the better predictor. Hayes thinks that's because machismo plays a role in the chili eating of men, but not women. "For women, there's no social status to being able to eat the hottest chili pepper, while for men there is," he speculates. Without the heavy hand of machismo on the scales, women's chili eating is more strongly governed by their internal drive for excitement.

Incidentally, while chili lovers laud the rush they get from a spicy dish, and sometimes claim the peppers "wake up" their palate to other flavors, you'll often hear chili-averse people complain that the burn keeps them from savoring other flavors in their meal. Which is it? The matter has received surprisingly little scientific study, but the bottom line seems to be that if capsaicin blocks other flavors, the effect is small. Most likely, when people complain that they "can't taste as well" after a spicy mouthful, it's largely because they're paying so much attention to the unfamiliar burn that the other flavors fly under the radar. In other words, it's not "hot" but "too hot" that interferes with the enjoyment of flavor—and the threshold where hot becomes too hot is a very personal one.

While chilis get most of the attention at the touch-related end of flavor, they aren't the only game in town. Of the others, one of the most intriguing is the tingling sensation from Szechuan pepper, a common ingredient in Chinese, Indian, and Nepalese cooking. If you haven't experienced this unique feeling, I encourage you to try it. You can find Szechuan peppercorns—which, despite the name, are neither chili nor black pepper, but flower buds of a member of the citrus family—in Asian groceries and specialty stores. They look like little brown Pac-men. Put a pinch in your mouth and chew for a moment, making sure it makes good contact with your tongue, and then wait a few minutes. At first, you might get a little hotness reminiscent of black pepper, but that's quickly replaced by a tingling sensation that's like nothing else you've ever experienced. Some people describe it as similar to touching your tongue to the terminals of a nine-volt battery. Others say it's like a vibration. "It's truly a crazy sensation," says Chris Simons, a food scientist at The Ohio State University who has studied it. "It doesn't hurt, it's not painful or irritating like capsaicin. You put it on your tongue and it actually buzzes." Specifically, the buzz is like a fifty-hertz vibration, as British researchers discovered when they asked people to match the feel of Szechuan pepper to a mechanical vibrator on their fingertip. If you have a piano handy, that's roughly the frequency of the lowest G, the seventh white key from the left.

The details aren't fully understood yet, but it looks like the active ingredient in Szechuan pepper, sanshool, blocks the flow of potassium out of nerve cells. This outward trickle of potassium acts to suppress nerve activity, so, in effect, sanshool tickles the nerves so that they're more likely to fire randomly. This case of neural jitters accounts for the buzzy feeling. Pharmaceutical companies are studying the same potassium channels as a target for painkilling drugs—and, in fact, after fifteen or twenty minutes, the buzz of sanshool gives way to a numbness that lasts for another quarter hour or so. This numbness can block some of the pain from chili peppers, Simons has found. Indeed, this may be one reason cooks began adding Szechuan pepper to their dishes, he speculates.

If your tastes run like mine, you're probably tempted to wash your Szechuan-pepper-laden *mapo* tofu down with a mug of cold beer. Good choice, because the fizzy bite of beer and soft drinks on your tongue is another common example of the mouthfeel side of flavor. If you've ever stopped to think about this sensation, you probably assumed that carbonation's bite is all about the bubbles. Until recently, most scientists thought so, too. "When I first started working with this, they said it's the popping of the bubbles on the tongue that makes the bite," says Bryant. But then Bryant chanced across a copy of a medical journal that made him rethink that simple story. The journal contained a letter from a doctor who was also a high-altitude climber. Like many such climbers, he took a drug to combat altitude sickness while on the mountain. On the particular climb in question, he had hauled a celebratory six-pack of beer up with him. When he cracked a bottle at the summit, he found that it had plenty of fizz but lacked the familiar bite. Intrigued, he and a colleague tested the effect back at sea level—

and sure enough, the altitude-sickness drug completely negated the bite of carbonation, even though the bubbles were still there.

The key to this puzzle is that altitude-sickness drugs inhibit an enzyme called carbonic anhydrase, which converts carbon dioxide—the gas that forms the bubbles in carbonated drinks—into carbonic acid. In the glass, CO_2 changes to carbonic acid very slowly, but once it gets into your mouth, carbonic anhydrase drives the reaction much faster. Since the drugs knock out carbonation's bite, that suggests that it's not the bubbles but the carbonic acid that is responsible for the bite.

Bryant and his colleague Paul Wise figured there was another way to test this idea, by keeping the CO_2 but eliminating the bubbles. "We took some seltzers and some beers into a hyperbaric chamber and cranked it up to 2 atmospheres," he recalls. The increased pressure kept the bubbles dissolved in the liquid, just as though they were still sealed in the bottle. "The seltzer had no bubbles and the same bite as when you drank it at normal pressure with bubbles." So much for popping bubbles. Instead, it seems that carbonation bite is all about the acid burn, yet another sensation largely detected by *TRPV1* receptors.

But something still nagged at Bryant, because he couldn't abandon the notion that bubbles were somehow part of the experience. So he and Wise tried another test, this time giving volunteers a mildly carbonated water, enough to generate a tiny bite but not enough to make noticeable bubbles. Then they slipped an aquarium air stone, the little, porous device used to aerate fish tanks, under the volunteers' tongues to add some bubbles—pure bubbles, no acid—to the experience. "We were essentially tickling the tongue with bubbles," he says. "And that increased the reported bite of the carbonation." It's not clear yet whether that's just because we expect the bubbles to be accompanied by a bite, or whether something else is going on as well.

So far, we've been talking about sensations that work much like smell and taste do, with nerve cells detecting the presence of particular chemicals—capsaicin, menthol, sanshool, acid—in a bite of food. The only difference is that in these cases, the information passes via touch nerves, rather than smell or taste nerves. But there are other layers in the mouthfeel sandwich, too, that are more a matter of touch in the usual sense—most notably, astringency. To appreciate this sensation, you just have to sip strong black tea or a tannic red wine—a young California cabernet, say—or eat an unripe banana. Recognize that dry, puckery feeling in your mouth? That's astringency. It happens when tannins and other compounds called phenolics in the food glom on to proteins in your saliva and prevent them from carrying out their usual lube job on your mouth and the food you're chewing. (If you put milk in your tea, your cup of tea is less astringent because the proteins in the milk tie up the phenolics before they can get to the salivary proteins.)

The perception of astringency may explain a lot about why certain foods go so well together. Think of red wine with steak; sorbet after rich, creamy soup; pickles with sausage; green tea with oily stir-fried pork and vegetables. Each of these pairs an astringent food or drink—often referred to as a "palate cleanser"—with one that's loaded with fat. Could fat and astringency be a culinary yin and yang, complementary opposites that bring out the best in each other?

The question intrigued Paul Breslin, the Monell researcher

who makes a habit of bringing his intense curiosity and enthusiasm to the dining table. A few years ago, Breslin and his colleagues decided to put this notion of palate cleansing to a rigorous test in the lab. To avoid all the complicated messiness of real foods, with their many ingredients, Breslin's team asked volunteers to sip standardized astringents—extracts of grape seeds or green tea—and describe the feeling in their mouths. The volunteers reported that the sensation of astringency built up over repeated sips, so that even a mildly astringent drink eventually became intensely puckery. But when the volunteers alternated bites of fatty dried meat with their sips of tea, each one kept the other in check. The fattiness of the meat tamed the tea's astringency, and the tea "cleansed" the palate of the meat's greasy feel. Remember that sip of wine the next time you grill a rib-eye steak.

Of course, now and then you find a food that manages to be both lubricating and astringent at the same time—chocolate being the prime example. That makes Hayes wonder whether delubrication is the whole story in astringency. If all that cocoa butter lubricating your mouth isn't enough to wipe out astringency, maybe we're detecting astringents more directly, too. Sure enough, German researchers—together with Linda Bartoshuk—recently reported the first hints that a receptor could also be involved in perceptions of astringency. The question remains open so far.

The fatty end of the equation, on the other hand, seems to be purely a matter of texture. As we've seen, our sense of taste picks up the nasty, rancid fatty acid part of fats, but not the rich, creamy, luscious whole. Instead, when we savor a buttery sauce or a bowl of ice cream, we're detecting the fat merely as a smooth, viscous coating in the mouth, a sensation picked up by ordinary touch receptors on the tongue and lips.

At this point we start to move beyond the so-called chemical senses of taste, smell, and receptor-based somatosensory perceptions like chili burn. In this wider world, our ordinary senses of touch, vision, and hearing have important roles to play as well. Just think of the difference between a crunchy potato chip and a soggy one, or a broccoli spear cooked just al dente versus one that's overcooked to mush. "Of the foods I like the most, certainly taste and smell are central, but on equal footing are things like lubrication, crunch, and chewiness," says Breslin. "If you think about how important texture and fattiness and creaminess and lubrication and even crunchiness are to what we eat, if you took those things away, it just wouldn't be enough."

Most people consider these textural attributes to be something separate from a food's flavor. I know I thought that way at the outset. But once these sensations get to the brain they're not so different after all.

Chapter 4

THIS IS YOUR BRAIN ON WINE

Among the countless restaurants that have come and gone on Upper Street in the trendy Islington district of London, the House of Wolf may have been the strangest. The nondescript, three-story gray brick building with a large mullioned window on the ground floor was once a Victorian music hall. Today, it has morphed into a nightclub called The Dolls House, but a few years earlier, it was what its owners called "a multi-functional, multi-sensory pleasure palace, dedicated to the creative pursuits of dining, drinking, art and entertainment." The rest of us called it London's most experimental restaurant. Its kitchen featured a predictably unpredictable parade of avant-garde guest chefs, each holding the stove for just a month or two before yielding to the next visitor.

If you had been lucky enough to stumble on the House of Wolf when it first opened in October 2012, you would have experienced one of the most peculiar meals of your life. As you enter the dining room, you're greeted by the sight of bread rolls dangling by strings from helium balloons floating at the ceiling. The chef—

artist Caroline Hobkinson—instructs you to put in earplugs, then eat your roll off the string without using your hands. As you nibble away—rather like bobbing for apples in midair—you hear the crunch of the crust, magnified by the plugs in your ears. "Can you hear the taste?" Hobkinson asks in the printed menu.

For the next course, you don a blindfold. Your waiter brings you a cracker topped with warm goat cheese redolent of rosemary and roasted red peppers. After your first bite, you remove the blindfold and see that the rosemary and pepper aren't on the cracker at all—they had merely been wafted before your nose as you ate the cracker and unflavored cheese. "Can you see the taste?" Hobkinson's menu asks.

No blindfolds or earplugs for the next course. Instead, your waiter sets before you a plate of salmon sashimi and a syringe filled with an amber liquid. Following instructions, you inject the fish with the liquid, which turns out to be ten-year-old Ardbeg, an intensely peaty Scotch whisky. Magically, the aroma of peat smoke from the whisky transforms the flavor from raw fish into smoked salmon. "Can you smell the taste?" the menu asks.

After a palate cleanser—a gin-infused cucumber ice, which you eat alternately with two spoons, one coated with salt crystals and the other with rose water crystals, giving two odd textures to your tongue—comes the main course, a straightforward, classic loin of venison with mushrooms, prunes, and wild cherries. *Ah*, you think, *something normal at last*. Well, not quite. Instead of a fork, the waiter brings you a tree branch as long as your arm, with the thick end carved into a forked prong. Like a Stone-Age hunter, you spear the meat with the branch and bring it to your mouth. "Can you feel the taste?" Hobkinson asks.

The final course, dessert, is a "sonic cake pop," a spherical chocolate brownie on a lollipop stick. It is served with an unusual garnish: a telephone number. You pull out your cell phone, dial the number, and hear instructions to press "1" for bitter or "2" for sweet. Depending on your choice, you'll hear a low rumble or a high whine—and the sound makes the dessert taste either bitter or sweet. "Can you dial a taste?" Hobkinson asks.

It all sounds a bit over the top—more performance art than meal. And at one level, of course, it is. But as with most art, there's a deeper message here, and Hobkinson is doing much more than just playing with our preconceptions about the dining experience. Her eccentric banquet also draws on a lot of solid science as it stretches our concept of flavor to include sight, sound, touch, and even thought. In fact, perceptual scientists can make a strong argument that a food's flavor isn't really contained in the food at all. Instead, you construct flavor in your mind from the whole range of senses you experience with each bite—and each of the courses in Hobkinson's meal is carefully designed to illustrate some part of that creative process.

Hobkinson's behind-the-scenes collaborator in all this is Charles Spence, a psychologist at Oxford University. A well-built man with a receding head of wispy hair, a deeply cleft chin, and a slightly protruding lower lip, Spence has the enthusiastic, self-satisfied air of someone who loves his work. And why wouldn't he? As one of the world's leading experts on what he calls "multisensory perception," Spence is forever playing with his food to better understand why things taste the way they do and how chefs, industrial food companies, and ordinary home cooks can heighten the flavor of the food they prepare. Along the way, Spence has collaborated with some of the best chefs in the world, including England's Heston Blumenthal and Spain's Ferran Adrià. Spence is one of the few scientists with the clout to command a VIP table at almost any high-end restaurant in the world.

Like many star scientists, Spence took a backdoor route into the research that made him famous. He was always interested in multisensory perception, but at first he focused on the better-studied senses of sight, sound, and touch. Back then, in the 1990s, hardly anyone worked on such "minor" senses as taste and smell. "It seems very bizarre, but most psychologists have only been interested in the so-called higher senses," he says. "There isn't much to read about food and flavor." But early on, he landed a few grants from food companies such as Unilever to apply the multisensory approach to flavor. Soon he was hooked, for both personal and professional reasons. "Food and drink are among life's most enjoyable activities, and they are the most multisensory, as well. It's an obvious place for a psychologist to end up," he says.

That makes researchers' lack of interest rather surprising—especially since, if you think about it in the right way, we all know that multiple senses must contribute to flavor. Imagine, for example, the wonderful taste of ripe strawberries slathered with vanilla-laced whipped cream. Easy, right? But now remember that all you really *taste* is sweet, plus maybe a little sour from the berries. All the rest is smell, experienced in the nose—yet it seems for all the world to be a taste, experienced in the mouth. Even worse, we often say the berries *smell* sweet, even though sweet is the one part of the flavor experience that we're not actually smelling. We've grown so used to combining smell and taste into a single flavor, though, that we commonly confuse them. "Maybe it's so common, in fact, that people never think about why strawberries smell sweet," says Spence. This kind of sensory magic also explains how the mere scent of peppers or rosemary lent their flavor to the plain goat cheese in Hobkinson's dinner—as long as the guests' dominating visual sense wasn't present to spoil the illusion.

Scientists, of course, are rarely content with this sort of vague hand waving, so Spence—along with several researchers in other labs—brought this sensory cross-wiring into the lab for dissection. More than a decade ago, for example, Richard Stevenson at the University of Sydney, Australia, had volunteers rate the sweetness of pure sucrose, an odorless sugar, both alone and in the presence of a caramel odor, which the researchers verified had no sweet taste on its own. Sure enough, the sugar tasted sweeter when people also smelled the caramel.

A whole host of similar studies show that the effect is widespread: Odors such as vanilla and strawberry also make sugar taste sweeter. Strawberry aroma enhances the sweetness of whipped cream, while peanut butter aroma doesn't. Chewing gum "tastes" less minty—really a smell and a mouthfeel, not a taste—as it loses its sweetness, and the mintiness returns when the researchers slip in a second dose of sugar.

Sometimes, these experiments point to another noteworthy fact: Smells and tastes often go together differently for different cultures. For example, caramel odor doesn't enhance sweet tastes for many Asian people, who are likely more used to encountering caramel in savory dishes instead of the sweets that Westerners are used to. The same thing happens with benzaldehyde, the main component of almond aroma. It enhances sweet tastes in Westerners, who usually encounter almond in pastries. But for Japanese, benzaldehyde enhances umami taste, because almond is a common ingredient in savory pickles.

In fact, researchers have found that they can mess with people's smell/taste perceptions almost at will. Several years ago,

John Prescott of the University of Otago, New Zealand, and his colleagues got their hands on some obscure odors that people would have no prior associations with. Then they presented those odors to volunteers together with either a sweet or a sour taste. After familiarizing the volunteers with the combination a few times, they tested the smells and tastes separately. Sure enough, sweet tastes seemed sweeter—and sour ones sourer—when the volunteers smelled whichever odor they had learned to associate with that taste. In short, we learn how to put smells and tastes together to create flavor.

It doesn't take a great leap of faith to accept that smell plays a big part in our perception of flavor. As Spence likes to point out, that notion has even penetrated that bastion of bureaucratic bean counting known as the International Organization for Standardization. As its name suggests, this is the agency that sets definitions and industrial standards for everything from telephone dialing codes (ISO 3166) to energy-efficient buildings (ISO 16818). If you like that sort of thing, and have the patience to wade through volume after volume of technical specifications, you'll eventually come across ISO 5492, which defines flavor as "a complex combination of the olfactory, gustatory and trigeminal sensations perceived during tasting." In lay terms, smell + taste + mouthfeel = flavor.

But that simple equation leaves out some crucial dimensions of flavor—dimensions that Hobkinson draws on in her multisensory feast, and that Spence has built a career around. A decade ago, for example, Spence did some of the pioneering work to show that our sense of hearing also contributes to flavor. In short, a steak's sizzle is part of its flavor.

Spence didn't actually use steaks in his experiments—they're expensive, and it's difficult to standardize "sizzle" in the laboratory. Instead, he turned to a foodstuff that could have been designed expressly with experimental psychologists in mind: Pringles potato chips. Instead of being sliced from individual, idiosyncratically flawed potatoes, Pringles are formed from a uniform slurry of pulped starch (rice, wheat, corn—and, yes, potato), so every chip in every can is identical to the next one—a perfect, standardized experimental replicate.

Spence and his associate Max Zampini asked twenty volunteers to munch their way through 180 Pringles each, rating the flavor of each chip, while wearing audio headphones that played back the sound of their crunching. As the volunteers soldiered on, a computer modified the playback sounds to make them quieter or louder, and to emphasize certain audio frequencies. The crunch, Spence found, was a key part of the chips' flavor. When volunteers heard a louder crunch, or even just a louder high-frequency part of the crunch, they rated the chips about 15 percent crunchier and fresher tasting than when they heard quieter sounds. The finding was surprising—and amusing—enough to net Spence and Zampini an Ig Nobel Prize, a tongue-in-cheek research award for research that "first makes you laugh, then makes you think." It's a laurel that Spence continues to wear proudly, mentioning it often in his later scientific papers, and even listing it among his "Academic Distinctions" right at the top of his résumé.

The same principle holds for other foods that feature distinctive sounds. Recently, for example, one group of researchers had volunteers rate the flavor of several coffees as they listened to sounds of a coffee maker in the background. Unknown to the tasters, every cup of coffee was actually identical—yet they rated the coffee 10 percent tastier when they heard sounds of a more "expensive" coffee maker (actually the same recording with annoying high frequencies muted).

Probably the best known of these experiments, in the foodie world, at least, was a test Spence conducted with oysters. He asked eaters to rate the flavor of the oysters while listening through headphones to one of two soundtracks: either sea sounds such as crashing waves and shrieking seagulls, or barnyard sounds such as clucking chickens and mooing cattle. By now you won't be surprised to learn that the eaters found the oysters tastier—and many people also found them saltier—when accompanied by the sea sounds.

This experiment has found its way straight onto the plate at what many people consider the finest restaurant in the world: The Fat Duck, in the little village of Bray, England, not far west of London. You'll find it just a few miles past Heathrow Airport, near the manicured grounds of Windsor Castle, where the queen likes to spend her weekends, and the famous playing fields of Eton, home to centuries of upper-class schoolboys.

It's not easy to get a table at The Fat Duck. Unless you can talk your way into a VIP table, you'll need to call at the stroke of noon, UK time, on the first Wednesday of the third month before your intended date. If you're lucky enough to score a reservation, expect to pay £255 (approximately $337, at the time of this writing) each, not including wine or tip. It's a small fortune, but you'll also spend four and a half hours savoring one of the most extraordinary meals of your life. Your menu might include an egg-white puff flavored with gin and tonic and frozen in liquid nitrogen, a quail jelly served alongside a bed of moss that emits forest-scented smoke as you eat, and something rather unpromisingly called "snail porridge."

But perhaps the most famous of chef Heston Blumenthal's creations is a dish he calls "Sound of the Sea." Your server sets before you a conch shell with a set of earbuds emerging from its opening. You put the buds in your ears and hear a soundtrack of crashing surf and calling gulls. Soon an edible seaside diorama arrives, with raw fish, seaweed, a seawater foam, and "sand" made from ground-up fish, seaweed, breadcrumbs, and other binders. Just as Spence found with his oyster experiment, the seaside sounds you hear as you eat are not just background frills but an integral part of the flavor experience. You're tasting not just with your mouth but with your ears as well.

Even abstract sounds can affect the flavors we perceive, as shown by Hobkinson's "sonic pop" dessert, where low-pitched sounds bring out the chocolate's bitter notes and high-pitched ones accentuate its sweetness, for reasons Spence cannot yet explain. Words, too, have "flavors," he's found—people associate spiky-sounding words like *kiki* with bitter flavors and rounder-sounding words like *bouba* with sweet ones. Other researchers have shown that people expect a mythical ice cream called "Frosh" to taste richer and creamier than one named "Frish."

If squeals, rumbles, and words can alter flavor, the logical next step would be to ask whether music can, too—and the answer appears to be that it does. "Sound is the last sense that people think about when it comes to flavor," says Spence, "but there's a huge explosion of work showing that people match flavors to classes of instruments or pieces of music." Heavy, powerful music such as Carl Orff's "Carmina Burana" makes tasters notice the heavy flavors in red wine, for example, while "zingy" pop music such as Nouvelle Vague's "I Just Can't Get Enough" brings out the brighter flavors in white wine, he notes. A few food writers are already working on cookbooks that pair food and music in "musical recipes"—and Spence himself says he now gives a lot more thought to his choice of background music for dinner parties.

He's also paying more attention to the crockery he serves his meals on, thanks to some other research he's done. As usual, this involved some trickery. In this case, Spence's colleague Betina Piqueras-Fiszman asked fifty volunteers to evaluate three different yogurts, presented one at a time in apparently identical bowls. By now you probably see the trick coming: In reality, the three yogurts were all the same, but Piqueras-Fiszman had invisibly weighted some of the bowls to be heavier than others. Sure enough, the raters judged the yogurt in the heavier bowls as being both richer and more pleasurable than the identical yogurt served in a lightweight bowl.

Even the color of the crockery can make a difference to flavor, Spence finds. In one test, for example, tasters rated a strawberry mousse as being sweeter when served on a white plate than on a black one. Most likely, he thinks, that's just because the white plate shows off the bright red strawberry color more dramatically, and this ripe-fruit color triggers us to expect sweetness. It's a simple effect, but hard to escape. As a result, says Spence, "I guess the black plates we used to serve on we no longer use."

Hobkinson was aiming for something similar in her dinner. The look and feel of forks carved from tree branches would subconsciously evoke associations with wildness, she hoped, thus enhancing the flavor of the venison. In effect, it's a kind of visual and tactile rhyme intended to emphasize a flavor message, just as a poet's rhyme emphasizes a verbal message.

That kind of visual rhyme turns up over and over in the world of flavor, and it usually works by modifying our expectations.

One study, for example, showed that a food's color can profoundly affect our perception of the food's sweetness, but not its saltiness. Presumably that's because in the natural world, color signals whether a fruit is ripe and sweet or underripe and sour, but we have no similar color clues to saltiness.

One experiment performed more than a decade ago—and now notorious among wine aficionados—showed just how powerfully our visually produced expectations can affect flavor, even for highly trained tasters. In this case, the victims were a set of budding wine professionals, fifty-four undergraduate students in the highly regarded enology program at the University of Bordeaux, France. One day, the students were given three glasses of wine—two red, one white—and asked to describe each wine's aroma. For enology students, of course, this is a routine task, and the students set about it with their usual thoroughness, discerning familiar aromas such as raspberry, clove, and pepper in the two reds and honey, lemon, and lychee in the white.

Gotcha! What the students didn't know is that there were, in fact, only two wines in the test, one red and one white. The third glass, the other "red," contained the same white wine, but researcher Gil Morrot and his colleagues had tinted it red with odorless food coloring. Simply changing the wine's color had totally altered the students' experience of the flavor. And remember, these were not naive, beer-swilling philistines, but people who were training for careers in the wine industry. (Their training might even have made them more prone to fall for the trick, because as experienced wine drinkers, they would have had stronger expectations linking color to certain flavors.)

So far, we've talked as though these multisensory effects on flavor act by altering our expectations, and there's no doubt that accounts for part of the effect. "When I smell a certain smell like strawberry, there's an expectation that what's coming next is going to be sweet," says Spence. Similarly, we expect "red" wine to smell like red wine, yogurt in a heavier bowl to be richer and more satisfying, red foods to taste sweeter than green ones. And what we expect to find, we do.

In this sense, everything about a meal's context—the paintings on the walls, the lighting, the tablecloths, and more—helps to create an expectation of the meal to come, and this expectation probably biases our perception of its flavor. To illustrate this, Spence and his colleagues recently held a public event in London's trendy Soho district where participants compared the experience of sipping a single Scotch whisky (The Singleton, for you Scotch geeks out there) in three different rooms. In the Nose Room, a green-lit space filled with leafy plants and the fragrance of cut grass, participants found that the whisky had a more pronounced grassy flavor. In the Taste Room, with red lights, rounded furnishings, and fruity aromas, the Scotch tasted sweeter. And in the Finish Room, a dimly lit, wood-paneled chamber redolent of cedar, its woody aftertaste became more prominent. All this, despite every participant knowing it was exactly the same whisky each time, because the cup never left their hand. Even circumstances that have nothing to do with food can bias our flavor perceptions. One study, for example, found that fans attending ice hockey games at Cornell University thought ice cream tasted sweeter after the home team won, and sourer after it lost!

This "everything contributes to flavor" attitude is not a new one. The Italian Futurist movement of the 1930s famously took the notion and ran with it to bizarre extremes. Diners at the movement's flagship restaurant, La Taverna del Santo Palato (the

Holy Palate Tavern) in Turin, dined on olives and fennel hearts with their right hands (sans cutlery) while stroking sandpaper and velvet with their left—all while the headwaiter doused them with perfume. Another course featured a sea of raw egg yolks surrounding a meringue island and airplane-shaped slices of truffle. I'm not entirely sure what expectations the futurist chef was trying to create, but suffice it to say that it didn't revolutionize Italian cuisine for very long.

But these multisensory influences on flavor aren't just a question of expectations. Even the faintest whiff of strawberry aroma—so faint it can't be consciously detected—is enough to boost our perception of sweetness, several studies have found. If you can't consciously smell the strawberry, you can't consciously expect a sweeter taste. Instead, Spence thinks, what's happening is something he calls "sensory integration." At first pass, that sounds not much different, but as Spence points out, expectations have a cause-and-effect timing pattern, in which you first smell the strawberry and then expect to taste the sweetness. In sensory integration, on the other hand, the two sensations arrive at the same time and reinforce each other.

You've experienced this kind of integration if you've ever watched someone's lips to help you hear what they were saying at a noisy cocktail party. Even when neither hearing nor lip-reading alone would be enough to understand the conversation, you can do just fine with both together. The simultaneity of sight and sound are crucial to doing this. "If you had the lip movements presented a half second ahead of the voice, the effect would disappear," Spence says.

This integration helps in understanding one of the great mysteries of flavor science, which you can experience for yourself right now. Take a bite or a sip of something flavorful—a rich stew, a ripe peach, a full-bodied wine—and take a moment to really savor it. Now, quickly, point to where the flavor is. Unless you're very unusual, you pointed to your mouth, but you know from chapter 3 that most flavor comes from your sense of smell, which is in the nose. The illusion is so strong that even knowing this reality, as you now do, doesn't change what you experience. So why does the flavor seem to happen in your mouth?

Sensory neuroscientists actually spend a fair bit of time obsessing over questions like this, and over the years they've arrived at a plausible answer. One of the brain's important jobs is to edit the raw stream of sensations, choose the relevant ones, and package them into concepts that we can think about. The timing of events is a valuable clue to that packaging: If two sensations occur together, they probably belong together. Ventriloquists exploit this tendency in their performances, by carefully timing their dummy's lip movements to match the sounds of their speech. If the match is good enough, the audience's minds will bind sight and sound together, leaving the strong illusion that the sound is coming from the dummy, not the human.

The same thing happens when you eat a mouthful of food. The bite delivers a whole suite of sensations—tastes, smells, texture, temperature, perhaps some crunch—all at once. The mind packages them together into a single experience, and assigns that experience to the mouth, where the most prominent physical stimuli occur. No one notices that some of the sensations—notably the food's aroma and the sound of its crunch—actually come from somewhere else.

With their penchant for taking things apart to see how they work in detail, neuroscientists have explored this further, of course. One particularly squirm-inducing experiment involved threading volunteers with a series of plastic tubes so that researchers could puff smells either through the nostrils or up the back of the throat, at the same time that they squirted scentless, odorless water into the mouth. The subjects identified the front-of-nose scents as smells coming from the outside world, but called the back-of-nose smells "tastes," and perceived them in the mouth.

We've just seen how the brain binds together sensations that come in at the same time, treating them as a single, unified flavor that can be greater than the sum of its parts. But as it turns out, not every set of simultaneous sensations gets bound into a unified perception. "In order to combine as a flavor, they need to be viewed as similar, as things that go together," says Johan Lundström, a sensory researcher at the Monell Chemical Senses Center.

As an example, Lundström, who's Swedish, points to an unpleasant experience common in kitchens back home. In Sweden, as in much of Europe, milk is often sold in cardboard cartons that cannot be resealed tightly after they're opened. When he happens to store an opened carton in the fridge with a leftover half of an onion, the milk picks up oniony odors—and that sensory dissonance makes it impossible for him to drink the rest, even when he knows it's fresh. "You cannot make yourself drink the milk," he says. "Your system is screaming at you that there's something wrong here."

That is healthy caution. One of the reasons we smell and taste in the first place is to make sure we don't eat something we shouldn't, so most people react with dislike to new or peculiar flavors, especially if they come as a surprise. (Adventurous foodies who know what they're getting into have other coping mechanisms that can override this aversion, says Lundström.)

This "go togetherness" turns out to be an important part of sensory integration. Lundström recounts an as-yet unpublished experiment that some of his colleagues at Monell did to test whether the brain is better at integrating flavors that go together—and whether we can teach our brains new flavor combinations.

Paul Breslin, Pam Dalton, and their collaborators made up special chewing gum that allowed them to give tasters tiny, measured doses of an aroma (rose scent) and either a bitter or a sweet taste. First, they figured out the minimum dose that people could detect of each scent or taste on its own, then dialed back the amount even more so that they ended up with what should have been flavorless gum. Then they paired taste and aroma in either familiar (rose–sweet) or unfamiliar (rose–bitter) combination gums. Sure enough, they found that people could taste the flavor in the gum with the familiar rose–sweet combination, indicating that they could integrate these two components, just like lip-reading at a noisy party. The people still couldn't taste anything in the rose–bitter gum, though, which shows their brains didn't know how to combine these discordant flavor stimuli into one integrated perception.

But then Breslin and Dalton took their experiment one step further: They made a new rose–bitter gum, but this time with enough of each flavoring that chewers could taste it. Volunteers chewed this peculiar gum daily for a month, then returned to the lab to see if their experience had changed their ability to taste the original, low-dose gum. And indeed it had—proof that after a month of practice, their brains had learned to integrate rose and bitter just as they had once automatically combined rose and sweet. "You can definitely train the system that these go together in a relatively short time," says Lundström.

All this evidence—from the gimmickry of Hobkinson's strange dinner to Lundström's oniony milk to Spence's Pringles and barnyard oysters—suggests that flavor is not what we usually think it is. Gordon Shepherd puts it best: "A common misconception is that the foods contain the flavors," he says. "Foods do contain the flavor molecules, but the flavors of those molecules are actually created by our brains."

The notion that flavor lives in the mind, not the food—or even the mouth or nose—is startling enough. But it looks like we can even go a step further: We construct our notion of flavor almost from scratch, building it from the ground up as we experience the world. Sure, a few preferences do appear to be hardwired, such as a liking for sweet tastes and an avoidance of bitter ones. But even those can be overridden by experience, as anyone who drinks gin and tonics can attest. And once we get to more complex flavors, it's clear that most of our perceptions and preferences are based on experience. To really understand how our brains create our experience of flavor, we need to dig into the details of what happens in the brain as we taste. First, a little background.

Psychologists generally think of the brain much like a layer cake. The bottom layer is made up of raw sensations—tastes, smells, touch, and so on. On top of that is a layer of synthetic perceptions, where raw sensations are assembled into objects: a series of shapes, colors, and shadings become a face, for example. Crowning the cake are one or more "cognitive" layers—exactly how many is a matter of debate—where higher-order thought takes place. Here, for example, we attach a name to the face and develop expectations of how that person will behave, how important they are to us, and so forth. For flavors, these cognitive layers are responsible for identifying and naming flavors, deciding whether they're good or bad, and choosing whether to eat something or not.

In this standard picture, all the information flows upward, with lower levels serving as data for higher processes. If there's no reverse flow, we'd expect the lower levels, the sensations and perceptions, to be "clean"—that is, driven purely by the sensory inputs themselves, and unaffected by any preexisting cognitive or emotional baggage. But we've already seen that's not exactly true, since experience can modify the way we bind sensations together. So what's going on here?

That's the question Edmund Rolls, a neuroscientist then at Oxford University, set out to answer. Rolls is one of the grand old men of sensory neuroscience, and many research paths in the subject led through his lab at one time or another. Rolls got to thinking about the particularly pungent spoiled-milk product that we know as cheese. Most people from Western nations like the stuff, while many Asians find it disgusting. (The tables are turned, of course, when it comes to Asian delicacies like aged duck eggs or the slimy, rotten-soybean preparation the Japanese call *natto*.) We know that cultural experience affects our liking for these foods. But, Rolls wondered, could these cognitive-level concepts reach back down and modify the raw perceptions, too?

To find out, Rolls and his student Ivan de Araujo devised yet another bit of psychology lab trickery. They prepared a synthetic "cheese flavor" and gave it to volunteers to smell. Half the volunteers read a label describing the odor as "Cheddar cheese," while the other half saw it labeled as "body odor." If you've been following along to this point, you won't be surprised to learn that the first group liked the smell better than the second group.

But then Rolls and de Araujo dug one step deeper, using brain scans to peer into the subjects' brains. There, they did find a surprise: the two groups' brains lit up differently all the way down to the second layer of the cake, the regions responsible for basic perceptions, even though nothing had changed except the words that described the odor. In other words, higher-level thought processes—and it's hard to find a level much higher than language—can change not just how we think about flavor perceptions, they can change the perceptions themselves. Thought itself, in other words, is one of our flavor senses. The brain constructs flavor by piecing together inputs from virtually every one of our sensory channels, plus inputs from thought, language, and a host of other high-level processes like mood, emotion, and expectation. That makes flavor a remarkably complex and changeable concept. It's a wonder we can talk about it coherently at all.

Actually, maybe we can't. Perhaps our flavor perceptions are so individual, so idiosyncratic, so circumstance dependent that we're fooling ourselves when we think we're saying anything objective about flavor. That's certainly the impression you get when you look more closely at wine. Wine should be a perfect test bed for exploring the reliability of our flavor perceptions. No other foodstuff is so thoroughly, obsessively described and quantified. Detailed tasting notes are available—usually from not just one, but several trained, professional tasters—for almost any wine available commercially. Not only that, but those tasters often assign numeric scores to every wine, too, allowing for numeric comparisons of quality. Wine, if you have the right mind-set, is where the world of food meets Big Data.

Bob Hodgson has the right mind-set. An oceanographer by training (now retired), he's also owned a winery in northern California for forty years. Like any other professional winemaker, he enters his wines in competitions such as the California State Fair, where trained judges taste their way through hundreds of wines and hand out coveted gold medals to the best—medals that can make or break a wine's salability on store shelves. Sometimes Hodgson's wines won gold medals, sometimes they didn't. But unlike most winemakers, he didn't just shrug at the injustice and carry on. With his scientific turn of mind, Hodgson started to wonder why the very same wine could garner a high score last week and a low one this week. Could you really trust the judges' scores, he wondered? Hodgson must be a persuasive guy, because somehow, he managed to convince the California State Fair to let him find out.

Judges at a big competition like the California State Fair taste about 150 wines every day, organized into 4 to 6 "flights" of 30 wines each. The wines within a flight are presented in identical glasses marked with identifying codes, so that no judge knows the identity of any wine he or she is tasting. Each judge individually—no discussion at this stage of the judging—gives each wine a numeric score on a 20-point scale. (Actually, the fair uses a 100-point scale like the ones you sometimes see on the shelves at your local wine shop. But any wine that's halfway drinkable scores at least 80 points, so for all practical purposes it's a 20-point scale.)

With the collaboration of the contest organizers—but unknown to the judges—Hodgson arranged that for one flight per day (usually the second), three of the thirty wines would actually be identical samples, poured from a single bottle of wine but given different code numbers. If judges' scores are a true reflection of a wine's quality, then you'd expect these triplicate samples ought to receive identical scores—or at least somewhat similar scores, allowing for a little bit of imprecision in the judges' ratings.

The results were shocking. "We did everything we could to make the task easy for the judge: same flight, same bottle. And nobody rated them all the same," says Hodgson. Only about 10 percent of the judges scored the three samples similarly enough that they awarded the same medal to each. Another 10 percent gave wildly different scores, giving one glass a gold and another a bronze or even no medal at all, and the rest fell somewhere in between. And that wasn't just because some judges are better than others: judges who were consistent in one year were no more likely to be consistent the next year.

Hodgson wasn't done. Next, he compared the results of wines that had been entered not just at the California State Fair but in other major wine competitions as well to see whether wines that aced one competition did well in others, too. As you can probably guess by now, they didn't. Wines would often win gold in one competition, nothing in another—and not a single wine out of more than twenty-four hundred picked up gold every time. The competitions might as well have handed out gold medals at random, Hodgson calculated.

So what's going on? The answer is that people's perception of a wine changes from moment to moment depending on the circumstances. The wine will taste blander if it follows a robust, fruity wine than if the previous wine was subtle; a particular aroma might have triggered a fond memory (and a higher score) for one glass but not the next one; the judge might have gotten tired as the flight progressed; they might have been distracted by a ray of sunlight or a twinge from an arthritic knee. All of that adds noise to the judge's rating—so much noise, Hodgson thinks, that it obliterates any real differences in quality. Maybe, in fact, it's just not humanly possible to judge wines objectively, especially in the crowded, rushed, overwhelming setting of a state fair.

Hodgson sees this variability at work when he drinks his own wine, too. "Since I have a winery, and I'm cheap, I drink my own wine all the time," he says. That's no hardship, because he generally thinks he makes excellent stuff. But even so, he's not always in the mood. "Sometimes I think, Jesus, I don't like this wine. But I know not to get upset, because tomorrow is a different day."

All this points to an uncomfortable conclusion: If trained judges and experienced winemakers don't consistently prefer one wine over another, then maybe there's no real basis for calling some wines great and others merely good. And that may be how it really is, though it's hard to find many wine people who will agree. "I would like to think that Mouton Rothschild is a better wine than Gallo Hearty Burgundy," says Hodgson. "You and I may agree that one is better—but we may not agree." Other studies, he notes, have found that ordinary wine drinkers, those of us without special training, tend to prefer cheaper wines over more expensive ones—but only if no one tells us what the price is. If you know the price, on the other hand, that high-level knowledge has a powerful influence on how you perceive the wine's flavor. Almost everyone tends to think a more expensive bottle of wine tastes better than cheap stuff—even when all that's changed is the price tag. That sounds like self-delusion—but there's more to it than that, as one team of researchers learned a few years ago.

A brain scan is not the ideal setting for savoring wine. For one thing, you have to hold your head perfectly still, which precludes all the sniffing, swirling, and other ceremony that usually accompanies a sip of wine. Instead, you get a tiny dollop—a single milliliter, about a quarter teaspoon—of wine dripped straight into your mouth through a polyethylene tube as you lie in the scanner. But at least researchers can see exactly what the wine's doing to your brain. For the experiment we're interested in here, the scannees got sips of what they thought were five different wines of varying price, but in fact four of the wines were paired duplicates: a five-buck plonk was also presented as a $45 bottle, and an exquisite $90 Napa cabernet also appeared under the guise of an everyday $10 wine. Sure enough, the tasters liked the wines better when they appeared with a higher price tag. But the brain scans showed that they weren't just saying so—the "higher-priced" wines activated the brain's reward circuitry more than the same wines presented at a lower price. In other words, a higher price tag genuinely led to greater pleasure! As one observer noted wryly, this means that if you're hosting a dinner party, you can maximize your guests' pleasure by serving them a cheap wine (which most drinkers prefer in a blind tasting) and telling them it's expensive.

As Rolls and other neuroscientists trace the flow of flavor through the brain, their attention comes back again and again to one particular spot, right behind the eyes at the front of the brain.

Neuroanatomists have a daunting catalog of tongue-twisting names for parts of the brain, most of which only an expert would need to know. But this little region, known as the orbitofrontal cortex, or OFC, deserves to be more widely known to anyone with an interest in flavor. The OFC, researchers are learning, is one of the key areas where the brain knits together the independent threads of taste, smell, texture, sight, and sound—together with our expectations—into the common cloth of a flavor perception. It's not stretching the facts to call the orbitofrontal cortex the birthplace of flavor.

(As is almost always the case with the brain the reality may be more complex. Another nearby brain region called the frontal operculum could also be a candidate for Flavor Central. In one recent study, researchers monitored brain activity while giving volunteers the odor or taste of orange juice either separately or together. The frontal operculum, but not the OFC, lit up far more strongly to the combined flavor stimulus than you'd predict from its response to smell or taste alone, suggesting that the frontal operculum may be another key area where flavor is constructed.)

If the OFC is where flavors are born, then it may also be the place to turn if you want to know what a flavor looks like in the brain. And, in fact, that's just what Rolls and his colleagues have done. By recording the electrical activity of individual nerve cells, or neurons, within the OFC of rats, they've found that each neuron there responds to a different set of inputs. One might light up in response to a sweet taste, a pepperlike aroma, and the mouthfeel of capsaicin, the molecule that makes chili peppers hot; another might turn on to sweet taste, a vanilla aroma, and the mouthfeel of fat. You could call the first cell a "chili pepper flavor" neuron and the second an "ice cream flavor" neuron.

This mapping of particular flavors onto individual neurons

helps explain why the first bite of, say, ice cream tastes so much better than the twentieth bite, and why we can eat our fill of stew yet still have room for pie. In essence, Rolls says, a particular flavor neuron gets tired after responding to its flavor over and over, a fatigue he calls "sensory-specific satiety." He's actually measured exactly this in monkey brains, showing how repeated doses of a particular flavor combination provoke smaller and smaller responses from its specific flavor neuron.

These flavor neurons in the OFC are also where learning enters the flavor picture. Remember Paul Breslin's rose–bitter chewing gum, which people eventually learned to treat as a coherent flavor? It turns out that Rolls had tried a very similar experiment in rats, while looking closely at individual neurons in the OFC. And indeed, when he switched up his odor/taste pairings, he saw those neurons gradually switch their responses to reflect the new associations. "You can watch the neurons learn," says Rolls.

While the neurons do relearn new odor/taste pairings, though, they aren't speedy about it. Rolls found he had to expose the rats to the new pairing about fifty times before they switched. In contrast, when he did the same experiment, except pairing tastes with visual signals instead of odors, the neurons began to relearn the very first time they saw the new pairing.

Why the difference? Well, says Rolls, it probably has to do with flavor's role in protecting us from eating the wrong things. "You don't want to realign your whole flavor system too rapidly." In the real world, smells tend to be rather reliably paired with the same tastes day after day, while the appearance of things can change quickly. Our brains, it seems, reflect this reality by being unusually conservative about taste/smell pairings, but looser with visual information.

Not that vision isn't crucial to our perception of flavor. Humans, after all, are a largely visual species, so it's not surprising that vision creeps into most of our experiences, says Lundström. A simple experiment shows how central vision is, he says. Try it: Imagine the fragrance of a ripe strawberry. Really focus on it. Now, didn't you also call to mind a mental picture of the fruit itself? "It's impossible to do that without actually visualizing a strawberry," says Lundström. "I think that vision is the key component when it comes to memorizing an odor, and you have a strong input from vision when it comes to odor quality."

To study this effect further, Lundström turned to a technique called transcranial magnetic stimulation—essentially an electromagnetic helmet that can be programmed to stimulate particular regions of the brain and make them work better. TMS of the visual center of the brain, for example, makes people about 10 percent better at discriminating among subtle shades of gray.

Lundström was after something more, though. If vision is linked to flavor processing, he wondered, could you stimulate the brain's visual system and improve flavor perception in the bargain? If so, that would tie vision even more firmly into the bundle of flavor senses.

So Lundström and his colleagues set up a study to test the idea. They offered volunteers three smell samples to sniff: two of one aroma and one of a different, but somewhat similar aroma—strawberry and raspberry, say, or pineapple and orange. The people had to identify which of the three was the different sample, a task they could do correctly about three-quarters of the time. Lundström had his subjects do the test three times: once without TMS, once with TMS stimulating their visual center, and once with a fake TMS helmet that emitted an impressive, official-sounding hum but caused no actual changes in the brain.

Sure enough, people treated with the real TMS, but not the pretend one, proved to be about 10 percent better at picking which odor was not like the others. In other words, helping people see better helped them to smell more accurately, too. And the effect wasn't just a general heightening of the senses. TMS of the visual center did nothing to help people tell which of three odor samples was more intense than the others, he found. That makes sense, says Lundström—visualizing the source of an odor is important in identifying it, but irrelevant in deciding how intense it is.

We've seen that the orbitofrontal cortex is the birthplace of flavor. It's also the crossroads for several other key parts of consciousness. All five senses pass through the OFC on their way into the brain, and the OFC also gets input from the brain regions responsible for emotions, reward, and motivation, as well as higher-order thought. The orbitofrontal cortex has been called the sensory packaging center of the brain, the place where all our world experience comes together. That suggests that flavor is not just a filigree on our lives, a little bit of aesthetic fluff. It's a key part of our interaction with the world.

Chapter 5

FEEDING YOUR HUNGER

Dana Small vividly remembers the first and last time she drank Malibu rum and 7UP. "It was a big party, and it was my first time, I think, having alcohol, and I was probably underage," she recalls. Small is, well, small, with long, bright-copper hair and a barely detectable soft lisp. "I didn't know what I was doing, and I probably didn't have that much, but Malibu and 7UP doesn't really taste like alcohol, it's like this really sweet kind of . . . Anyway, so I had a few of those, and felt not so good the next day. That was 20 years ago. In those 20 years, I've continued to eat a lot of sweet things, but I particularly avoid Malibu and 7UP."

Most of us can think of a similar event in our past, where a bad experience has permanently scarred our taste for some particular food or drink. But for Small, a neuroscientist at Yale University, the lesson runs much deeper than "Don't drink this." She thinks experiences like hers—and the positive ones on the flip side of the coin—are the whole reason our brains assemble a unified perception of flavor from what could have been left as separate senses of taste, retronasal smell, and texture. "The reason that we have flavor, when we already have taste and smell, is for the purpose of associating foods that we encounter in the environment with their post-ingestive effects, because ultimately that's what it's all about. That's really the role of flavor," she says. Translation: We remember the flavors of what we ate, and what happened afterward, so that the next time we can seek out the good stuff and avoid the bad. "Flavor perception allows us to have a representation of precisely a particular kind of food. So in the case of Malibu and 7UP, there is specific learning to avoid that item. This is really like no other kind of learning. It's very strong—one trial—and very long lasting. That makes perfect evolutionary sense: you want to only need one trial."

For our ancient ancestors, omnivore hunter-gatherers that they were, these eating decisions would have been far more than a minor matter of aesthetics. Their choices of what to eat could literally be a matter of life and death. Pick the wrong root to eat, and you poison yourself and your family. Pass up a nourishing root, and you could all starve. More subtly, being a successful hunter-gatherer hinges on finding the foods that deliver the biggest nutritive bang for your hunting, gathering, and chewing buck. In modern terms, if you're uncertain you're going to eat tomorrow, then today you damn sure want to eat potatoes or burgers or ice cream, or something else with a lot of calories, rather than wasting your time munching raw celery.

So you'd expect that evolution would have endowed us with a pretty good system for identifying and remembering potential foods and what happens when we eat them. A little bit of this system, as we've seen, is built-in: even newborns have an innate liking for sweet tastes. But mostly, we learn through experience. That's what flavor does for us—and why our brains assemble all the relevant data of taste, texture, retronasal olfaction, and all the rest into a single, unified flavor perception. Thanks to this synthetic perception, we can learn and remember the flavors that made us sick, and we also learn to like the flavors that nourished us. We don't generally notice that we're learning about nourishment, because in our everyday world, flavors and calories are inextricably linked. We don't usually encounter a baked-potato flavor without also ingesting a big slug of carbs, or salmon flavor without protein and fat. Teasing the flavor apart from its nutritional consequences takes careful experimentation, the sort that's easier to do with rats than with people.

The classic studies here come from Anthony Sclafani, a researcher at Brooklyn College in New York. Sclafani offered rats one of two water bottles to drink from, one grape flavored, the other cherry flavored. Neither contained any sweetener or other nutrients—just flavoring and water. But he also inserted a stomach tube, so that he could deliver sugar solution straight into the rat's gut when it drank the cherry flavor, but not the grape. Since the sugar never entered the rat's mouth, it tasted no sweetness. Yet within a few minutes of encountering the two flavors, the rat learned to drink almost exclusively from the cherry-flavored water bottle. What's happening, says Sclafani, is that nutrient receptors in the rat's gut quickly signal to the brain that good stuff is coming in. The brain pairs this with flavor information from the nose and mouth, and the rats learn that cherry means calories, even though they never taste the sweetness. Sclafani also reversed the pairings in other rats, just to be sure there wasn't something special about cherry flavor. Rats that got sugar infused into their stomachs when they drank the grape-flavored water quickly learned to prefer grape instead. And it's not just sweetness—Sclafani's rats learn just as well if the calories delivered through the stomach tube come from proteins or fats. It's exactly the same learning process that Russian biologist Ivan Pavlov used to train dogs to associate the ringing of a bell with imminent food. After not too long, Pavlov could just ring the bell and his dogs would start salivating in anticipation of the meal to come. Sclafani's rats hit the cherry-flavored water because they expect calories to follow.

By moving the location of the rats' stomach tube, Sclafani was able to show that the nutrient receptors responsible for the learning are located right at the beginning of the small intestine, just past the stomach. This is the region that surgeons remove when they do gastric bypass surgery on morbidly obese patients. No one knows exactly why gastric bypass surgery works so well, but one reason may be that it eliminates these nutrient receptors and thus prevents the pairing of flavors to their nutritive consequences. Since the flavors of a meal are no longer associated with incoming nutrients, people would gradually lose interest in the flavors and feel less drive to eat them, Sclafani suggests.

This flavor learning—technically known as flavor-nutrient conditioning—is dead easy to demonstrate in rats, as Sclafani's work shows and others have verified. But it's a lot harder to prove that the same kind of learning happens in people. For one thing, the whole stomach-tube business is a nonstarter. People also have an annoying habit of eating whenever they feel like it, so that experimenters have a much more difficult time controlling their food intake or ensuring that they haven't already formed associations with, say, grape and cherry flavors. As a result, studies of flavor-nutrient conditioning in humans have had mixed results. Sometimes it looks like it happens, and other times it doesn't.

Probably the best evidence that we really do learn to pair flavors with their nutritive rewards comes from Dana Small's lab at Yale. Small flipped through a flavor-supply catalog to come up with ten really obscure flavors that no normal person had much chance of encountering in daily life. "They're novel, and you don't know what they are," Small told me when I asked her to describe the flavors. One, for example, was called "aloe," though it tasted nothing like aloe vera. Not surprisingly, people tended not to like the flavors when they first encountered them—our built-in neophobia raising its head again.

Small and her colleagues picked two flavors and used them to make artificially sweetened soft drinks. One of the flavors also got a dose of maltodextrin, a sugar that delivers a full load of calories—it turns into glucose almost immediately when it reaches the stomach—but is devoid of flavor. (A triangle test—which of these three is not like the others?—confirmed that people couldn't tell the difference between soft drinks with and without the maltodextrin.) Volunteers consumed each drink several times over the course of a few days, using only one kind of drink each day to keep the postingestive consequences separate. And then Small brought them back into her lab to see how they responded to the two flavors. To be sure she was looking at the effects of learning, and not real-time perception, this time neither flavor was spiked with maltodextrin. Sure enough, the people showed a slight tendency to like the high-calorie flavor better than the low-calorie one. In other words, they had learned which flavor delivered the nutritive goods, and they liked it a little better—but not all that much better. The big difference showed up when Small put them in a brain scanner.

Being in Small's brain scanner is not exactly a fine dining experience. Just like any hospital MRI, you're flat on your back inside a giant magnet, with your head immobilized. To get a good image of the effect of each flavor on brain activity, she needs to average over multiple sips: on again, off again. She needs to know exactly when the flavor arrives, and she needs to keep stray odors from lingering and confusing the test. "What that means," Small says, "is that you've got a nasal mask, and then this teflon thing that liquids are dripping off onto your tongue." Charming.

Even in that utterly strange context, the results were dramatic. When people drank the flavor they'd learned to associate with calories, a part of the brain called the nucleus accumbens lit up like a Christmas tree. The nucleus accumbens is a part of what's often described as the "reward pathway," the part of the brain where good things begin to feel good, so that you want to do them again. The reward pathway plays a role in making you want more of things like sex, drugs, and rock and roll (literally—music activates the nucleus accumbens). An old study from the 1950s hooked rats up so they could stimulate their nucleus accumbens by pressing a lever; the rats just kept pressing the lever, over and over and over again, not even pausing to eat or drink.

Crucially, the learned flavor-nutrient link swayed the response of people's reward pathways much more strongly than it affected their conscious liking of the two flavors. Let's pause for a moment to underscore that point: When Small asked her subjects which flavor they preferred, she didn't find all that much difference. That might explain why previous studies of flavor-nutrient conditioning in humans haven't been very convincing. But Small didn't stop there. Instead, she also let the subjects' brains tell her which flavor they valued more—and their brains spoke loud and clear. All the real work, it turned out, was happening under the surface, in the unconscious.

Small points to another recent study that reinforces the point. Researchers at her alma mater—McGill University in Montreal—wanted to separate our conscious and unconscious valuations of food items to see how they differed. To do this, they showed pictures of food to hungry volunteers, and asked them to estimate their calorie content. (That's the conscious valuation.) At the same time, a brain scanner measured the activity in a region of the brain called the ventromedial prefrontal cortex, another area involved in valuation and appetite. (That's the unconscious valuation.) To top it off, the subjects were also given five dollars and asked how much of it they'd pay to have that food item to eat right now. Remember, these were hungry college students, who presumably cared about getting some calories at the time.

People turned out to be pretty lousy at consciously guessing how many calories the food items contained. Their unconscious brains, however, did much better: Their brain activity matched the real caloric content of the foods, not their estimate of the calories. The interesting result, though, showed up in people's willingness to pay for the food. You'd think that when people are consciously deciding how much to pay for a snack, they'd base their decision on their conscious estimate of calorie count. But in fact, the amount they paid was a much closer match to the actual calories—the information accurately assessed by their unconscious.

At this point, you might be wondering why people persist in drinking Diet Coke, or continue to put sugar substitutes in their coffee. You'd think that their bodies would learn that those flavors don't deliver calories and thus aren't worth craving. One reason we don't learn to ignore those flavors is that they deliver a jolt of caffeine, which also feels good. Our bodies learn to like the flavors associated with that kick—and with the buzz of alcohol, too.

That's why so many of us so easily develop a predilection for what are, objectively speaking, nasty, bitter, burning flavors.

There's another point to consider when it comes to fooling the flavor system. You might be a dedicated Diet Coke drinker, but you probably encounter some of the same sweet, citrusy, caramely flavors in other foods, too, where they're accompanied by real calories. That variability—sometimes sweet citrus means calories, sometimes not—might interfere with flavor-nutrient conditioning and make it harder for our internal calorie counter to keep track of how much we've eaten and when to stop. We might even be making matters worse, because we turn the flavors into a caloric slot machine that sometimes pays off and sometimes doesn't. This sort of "intermittent reinforcement," to use the technical lingo, is especially good at snaring our reward pathway. (Just look at all the zoned-out people sitting in front of actual slot machines in your nearest casino.) If so, artificial sweeteners might actually increase our attraction toward sweetness and the other flavors that accompany it. This may help explain why artificial sweeteners haven't exactly been a weight-loss panacea.

It makes good evolutionary sense that all of this sophisticated learning takes place below the threshold of consciousness. Long before human beings ever walked the planet, and even before the first primates picked their way through the trees looking for fruit, our primitive mammalian ancestors would have needed to identify which foods were most nutritious. In short, they would have needed flavor-nutrient conditioning. And they probably had little or no conscious thought to help them with the task. "These circuits evolved so long ago," says Small. "They were working perfectly well before we had consciousness." As good mammals, then, we have evolved to want the flavor of calories. Or, to put the matter more precisely, we want the flavors that we've learned are accompanied by a dose of calories, while we ignore the flavors that aren't. And this happens mostly without our conscious awareness.

But modern humans, with very few exceptions, no longer live on the African savannas, digging up roots and picking fruits and running down the occasional gazelle. We're surrounded by an abundance of foods, and many of them are calorie rich in a way our ancestors rarely experienced. In this new context, our evolved instincts let us down. We no longer benefit by being attracted to high-calorie flavors when they're always there—and the increased caloric density kicks our flavor-nutrient conditioning into overdrive, making those foods even more attractive. We want those flavors even when getting them is bad for us.

As we've seen, a few of our flavor preferences are clearly innate. Even newborn babies like sweet tastes—and they have to, otherwise they might not latch on to the mother's breast and feed. And those same babies naturally reject bitter tastes, which are often an indication that something is toxic. Once past those few simple cues, though, our flavor preferences are wide open. Each of us has to decide which potential foods we will eat and which ones we will shun. A panda doesn't have to learn this—it eats only bamboo. A lynx eats rabbits. An anteater eats ants. But people are different: as omnivores, we have to learn the flavors that mark the foods we eat.

That learning starts before birth, as flavor molecules from foods eaten by a pregnant mother pass into the amniotic fluid and are ingested by the developing fetus. In essence, the fetus samples what the mother eats—and, later on, recognizes and likes those flavors. Nursing infants get the same chance to sample mom's diet through breast milk. The best demonstration of this early learning comes from Julie Mennella at Monell. Mennella asked one group of pregnant women to drink a glass of carrot juice at least four days a week during the last trimester of their pregnancy. Another group drank the carrot juice not during pregnancy but while nursing their infants, while a third group never drank carrot juice. Later, after the infants had begun to try solid food, Mennella watched as they got their first taste of carrot-flavored baby food. Most babies make scrunchy faces when they taste something new, but the babies who tasted carrots in utero or while nursing made fewer expressions of distaste than babies whose mothers had avoided carrot juice during pregnancy. The carroty babies' mothers also thought they enjoyed the carrot-flavored cereal more. In short, the babies who'd experienced carrot through their mother were more comfortable with the flavor when they first encountered it directly.

And it's not just carrots. Over and over again, researchers have shown that babies who are exposed to flavors ranging from anise to garlic through the mother's diet prefer those foods when they first encounter them directly. In short, we learn to like what our mothers eat. "It's a really beautiful system," says Mennella. "For the baby to learn to like a food, the mother has to eat it. You can't pretend to eat it, because the flavors don't get in."

Those early, learned preferences can linger for many years. In Germany, for example, almost all infant formula used to be flavored with vanilla. Many years after the practice had stopped, researchers took advantage of this natural experiment by comparing taste preferences of children who were infants just before the change—and who therefore almost certainly drank vanilla-laced formula—with those born just a few years later who drank vanilla-free formula. Sure enough, the children who had tasted vanilla as infants liked it better, years later, than those who hadn't.

The odd bit of vanilla flavoring notwithstanding, formula-fed infants don't get the same exposure to the flavors of foods their mother eats. Instead, they get exactly the same set of flavors with every bottle, unless the parents switch brands now and then. A formula-fed infant, then, arrives at weaning with little or no experience of the flavors he or she will soon experience firsthand. That's especially true of infants who feed on cow milk formula or soy-based formula, which tend to be sweet and bland. In contrast, formulas made of hydrolyzed protein have bitter and sour flavor notes, so that infants who drink them get some familiarity with those "difficult" tastes. That familiarity helps those babies—like their breast-fed neighbors—be more accepting of vegetable flavors when they eat their first solid food, Mennella finds.

There seems to be a window of opportunity during the first few months of life when babies will accept almost anything they're given—even hydrolyzed protein formula, which most adults find nasty tasting. But even after weaning, toddlers and older children gradually learn to accept new foods if they try them often enough. Toddlers and young children tend to be wary of new flavors and will often reject a new food the first few times they sample it. After eight or ten tries, though, most kids will begin to accept it in their diet—although they'll often continue making skeptical faces well after that. (Parents should ignore the faces and pay attention to what their child actually eats, Mennella says.) Simple variety seems to matter, too, so children who have the opportunity to sample many different foods are more likely to accept new flavors. And they learn by watching what their parents and older siblings eat.

The lesson for parents is very clear: Eat what you'd like your children to eat. "Children learn through repeated exposure, variety, and modelling. I don't know what more I can say," says Mennella. "It's just the basic tenets of learning—and learning as a family. Foods identify what that family is. Eat the healthy foods that you enjoy and like. Offer them to your kids in a positive context. Kids will learn."

Perhaps the most vivid demonstration that it's easier to learn to like something if you grew up with it comes from the high Arctic. The Chukchi and Yupik people of the Bering Strait region traditionally lived on a diet of fish and walrus, and many favorite dishes involve burying meat, blood, and fat to ferment for months and become what the locals call "tastily rotten." To someone who hasn't grown up with the practice, such foods can be hard to take, even with all the open-minded goodwill in the world. Here's how one anthropologist, eager to try indigenous foods, described her first encounter with aged walrus meat:

> What a shock! The smell of the thoroughly aged meat permeated my senses. My only thought was that as a guest I should not be rude. I must finish this piece of meat. I chewed and chewed and chewed. . . . Finally [one of the hosts] said quietly, still smiling, "You know, Carol, you are turning green!"

Even among the Chukchi and Yupik themselves, their fondness for these foods depends on their childhood experiences. Elders raised on traditional diets generally love them and still go out of their way to eat them. However, a generation of children raised during the 1960s–1980s, when the Soviet government actively discouraged traditional foods, often struggle to eat these foods.

Their aversion was so strong that many still refused to eat them even when they had little choice, after outside food became very scarce after the collapse of the Soviet Union. And anthropologists report that many young people today are happy to honor and spend time with their grandparents—but only if they can leave before dinnertime. Even those who now eat these tastily rotten foods sometimes do so wearing latex gloves to avoid the lingering aroma.

A few preferences, in contrast, don't seem to be open to learning. Newborn babies suckle more and make happy faces when they have something sweet in their mouth. And there's not much that parents can do to change that preference. Gary Beauchamp—the same flavor researcher I gargled with in the Monell boardroom—tried to get people used to eating less sugar. Decades earlier, Beauchamp had found that putting people on a reduced-salt diet for a few weeks shifted their tastes so that they preferred the less salty food over what they used to eat, which they now regarded as too salty. But when Beauchamp tried to do the same experiment for sweetness, he found that people didn't respond the same way. After three months on a low-sugar diet, his test subjects preferred exactly the same sweetness of vanilla pudding or raspberry drink as people who ate their usual diet. To Beauchamp's knowledge, no one else has tried a similar study, and he cautions against concluding too much from a single experiment. However, if he's right—and if children respond the same way as adults—then parents might be able to relax a bit about sugar. "Every person on the face of the Earth, almost, believes that if you feed kids lots of sugar, they're going to like it more. There's no evidence of that," he says. Nor does a sugar-free diet keep kids from craving sweetness. (Beauchamp recalls one child he tested whose parents were fanatical about avoiding processed sugar and other sweets. The kid told him that at school, he got his sweet fix by pulling used chewing gum off of chairs and chewing that.) Sweet is sweet, and it always tastes good—and there's nothing parents can do about that.

In most societies today, the most pressing question about food is how we can nudge people to eat less of it. As everyone knows, Americans have been gaining weight for decades, and more than two-thirds of American adults now weigh enough to be classified as overweight or obese. Europeans and even Chinese and Indians are now starting to join them. Worldwide, 39 percent of adults are overweight or obese, and overnutrition now kills more people each year than undernutrition.

Since this book is about flavor, we're going to look at just one tiny piece of the puzzle: how the flavor of our food helps determine what we choose to eat and—more critically for weight control—how much of it we pack away. Even that little piece of the diet pie gets tricky. The first complication arises because we generally stop wanting something once we've had enough. The clearest illustration of this is the phenomenon of sensory-specific satiety that we touched on in the last chapter. The more you experience the flavors of a single food item during a meal, the less your brain's reward system responds to the sensory input, no matter how many calories it packs. Even if you really like something, you enjoy it less with each bite.

This is a big reason why high-end chefs gravitate toward tasting menus that feature a long progression of tiny dishes. At Chicago's Alinea restaurant, for example—generally rated among the world's best—you can expect a meal of more than a dozen tiny courses, each consisting of just a few bites. The chef, Grant Achatz, learned much of his craft at the renowned French Laundry in California's Napa Valley, another restaurant famous for serving a long succession of "small plates." Here's how Achatz's mentor there, chef Thomas Keller, described why he does that:

> Most chefs try to satisfy a customer's hunger in a short time with one or two dishes. They begin with something great. The initial bite is fabulous. The second bite is great. But by the third bite—with many more to come—the flavors begin to deaden, and the diner loses interest.

The same principle applies to any meal. If your holiday meal consisted of nothing but mashed potatoes, you'd almost certainly eat a lot less than you do when you've also got turkey, stuffing, green beans, and brussels sprouts on your plate. Sensory-specific satiety kicks in within fifteen to twenty minutes after you start eating a food, so it may help turn off your appetite for what's on your plate even before other satiety signals like stomach fullness kick in. (The effect wears off within an hour or so, so other satiety mechanisms must be more important in determining when we start thinking about eating again.)

Some people have suggested that more highly flavored foods might be better at inducing sensory-specific satiety. If so, people might be able to lose weight by maximizing the flavor per bite. For example, Edmund Rolls—who discovered sensory-specific satiety—notes that in most cultures, the staple foods, the ones that people eat the most of, tend to be relatively bland starches like rice, potatoes, or bread; the more highly flavored meats and vegetables usually take a smaller share of the plate. On the other hand, there's only a little evidence to date to back the notion that maximizing flavor could keep us thinner.

Perhaps the clearest support comes from a recent Dutch study where researchers used plastic tubes threaded up the back of volunteers' throats to deliver either a strong or a weak tomato soup aroma. Either way, the volunteers drank the same bland, flavorless soup, but they perceived it as either a richly flavored or mildly flavored tomato soup, depending on the intensity of the piped-in aroma. Sure enough, they ate about 9 percent less of the more highly flavored soup, as long as the more intense aromas were present more than fleetingly.

If sipping soup with a hose up your nose sounds unpleasant, just be thankful you didn't end up in the experiment that some other young Dutch men volunteered for a few years ago. Researchers wanted to sort out whether we stop eating because our stomachs are full or because we've had enough flavor. To answer that question, they needed to separate chewing and flavor release from swallowing and filling the stomach. The way they did it wasn't pretty.

If you'd been part of that experiment, you would have come in to the lab sometime after eating a normal breakfast, and a technician would have threaded a skinny stomach tube, thin enough to poke into the headphone jack on your phone, up your nostril, down the back of your throat, and into your stomach. After an hour of sitting around and filling out paperwork, you'd have been given a half pound of cake. You'd have been told to chew the cake normally, but then, just when you were about to swallow it, spit it out into a cup instead. Take another bite, chew and spit, over and over, for either one or eight minutes. When you were done, a technician—presumably the one who drew the short straw—collected the contents of the cup, dried them, and weighed them to make sure you hadn't swallowed any cake on the sly. While you chewed, the stomach tube flavorlessly deposited into your stomach ninety-nine calories' worth of the same cake, pureed in a blender with either a small amount of water (one hundred milliliters, which is around three ounces) or a larger, more belly-filling amount (eight hundred milliliters, around three cups). A half hour after all the spitting and pumping (and after the stomach tube was snaked out of your nose again), you would have been given a sandwich lunch and instructed to eat your fill.

Sounds miserable, all this chewing and spitting and slithering of tubes. And apparently, the study participants thought so, too. Of the forty-three young men who signed up for the job, eight walked away once they found out what was involved, even though they would have been paid for their time. Five were sent packing for failure to spit out all their cake, and four washed out for other reasons, leaving just twenty-six to finish the experiment.

After all that, it turned out that when it comes to satiating us, flavor in the mouth matters at least as much as fullness in the stomach. On the days when the volunteers spent eight minutes chewing and spitting out their cake, they later ate 10–14 percent less of their sandwiches than when they chewed and spat for just one minute or not at all. In contrast, their sandwich consumption barely went down at all on days when they had more fluid pumped into their stomachs.

In another, slightly less unpleasant experiment, the same researchers pumped tomato soup into people's mouths, either as big squirts separated by twelve-second pauses, or a series of small squirts separated by three-second pauses. Either way, the eaters got the same amount of soup per minute, but they got less flavor from the big gulp, because they spent less time with soup in their mouths. Sure enough, when people got the little doses—hence the more flavor exposure—they ate less soup before deciding they were full.

All of this would seem to suggest that advocates of thorough chewing may have a point: The more you chew, the more you're exposed to food flavors, and therefore the quicker satiety may set in. In one study, people felt fuller after eating pasta with a small spoon, dutifully chewing each mouthful twenty or thirty times, than when they used a large spoon and ate as quickly as they comfortably could. (Unfortunately, the science isn't quite clear-cut yet—though the long chewers felt fuller, they didn't actually eat any less of the pasta despite feeling more sated.) Even if you buy into the notion that chewing more means swallowing less, you don't have to take this to fanatic extremes—as advocated, most notoriously, by Horace Fletcher early in the twentieth century, who sparked a brief fad for "Fletcherizing," or chewing every mouthful hundreds of times. Instead, you may be able to achieve the same end without obsessing over chews by playing with texture. Foods that are thicker, chewier, or crunchier force eaters to take smaller bites and chew longer, which slows the eating rate and ups mouth time.

More to the point, liquids such as soft drinks, juices, and beer go down much more quickly than chew-and-swallow food—as much as ten times faster, according to some studies. We get less exposure to their flavor in the mouth, which may explain why we tend to overconsume liquid calories: They don't trigger our internal calorie meter as strongly as calories from solid foods do. And in fact, people find identical amounts of soup more filling if they eat it from a spoon, slowly, than if they drink it from a mug more quickly.

Of course, another way to experience more flavor from your food is simply to eat more flavorful food. No one knows whether tastier meals will make you feel full more quickly, though many of the experts I spoke to said they wouldn't be surprised if that was the case. And a few experiments hint at the possibility. People take smaller bites of more highly flavored vanilla custard, for example, and they ate less of a saltier tomato soup than of a less salty one that they rated equally desirable.

It's tempting to think that the immense variety of foods available today might contribute to overeating, because when sensory-specific satiety kicks in we can just switch to a different food. When obesity first appeared on the social issues radar, back in the 1970s, a lot of people worried that the root cause was our modern "cafeteria" diet, in which we're constantly exposed to a huge variety of potential foods. With so much to choose from, people worried, we'd hop from one food to the next and end up eating too much. It rapidly became clear that variety, per se, was not the culprit. In the early 1980s, researchers at Monell bought commercial flavors in a dozen rat-friendly flavors. (In case you ever want to charm a rat, the flavors were peanut, bread, beef, chocolate, nacho cheese, cheese paste, chicken, cheddar cheese, bacon, salami, vanilla, and liver.) Then they fed the rats standard rat chow spiked either with the same one flavor over and over again, or else a constantly changing smorgasbord of varied flavors. If variety causes overconsumption, the latter rats should have blown up like happy little blimps.

But they didn't. Over the three weeks of the experiment, the smorgasbord rats ate no more food, and gained no more weight, than the boring-food rats. What really mattered, it turned out, was how much sugar and fat was in the rats' diet. A third group of rats that got a much higher-fat, higher-sugar diet (let's call it the fast-food diet) did indeed balloon up, regardless of whether they got a diet with monotonous or varied flavor. In other words, it's not about the variety. It's about the reward-pathway pull of concentrated calories.

Even so, many people manage to navigate past the siren calls of fast-food calories without gaining weight. That's because we have a separate system for regulating the total amount of food we take in. A complex network of hormones with names like leptin, ghrelin, and neuropeptide Y regulates our levels of hunger and satiety to keep calorie input roughly equal to output in the long run. When I eat a varied holiday dinner, I'm likely to eat a little bit more than usual. Most people do—not just because of the variety, but also because the social context says we're supposed to. But we compensate for that later, by eating a little less the next day, or by skipping a snack or two. And you can forget the old saying, "Never trust a skinny chef"—as a general rule, people don't gain weight just because their diet is especially tasty. "There's nothing out there—and I've asked around—that really shouts that if you make food as palatable as you can, people overeat," says Mark Friedman, a researcher who's been studying flavor and appetite for decades. "For something that everybody believes, there's not much there."

Nor does bad-tasting food make you eat less. (Just ask any college student on a meal plan.) If you make rats' chow less palatable, such as by adding a bitterant to it, they'll avoid it for a little while, but if they have no other options, hunger kicks in eventually and they scarf the stuff down anyway. Likewise, people who suddenly lose their sense of smell—a malady we discuss later in this chapter—usually don't end up losing weight. For that matter, most of us can think of someone we know who cooks boringly, or even badly, and they're generally not emaciated.

Further evidence that we probably can't fix obesity by tinkering with flavor comes from the field of genetics. You'll recall that each of us carries a unique set of genetic variants in our taste and odor receptors, and as a result no two of us perceives the flavor world in exactly the same way. If our flavor perception was an important cause of obesity, then you'd expect that people with some flavor-gene variants would be more likely to be obese than others. For example, people with my sweet-receptor variant tend to prefer sweeter flavors, which might put us at greater risk of eating too many sugary treats and gaining weight. Or people who are especially sensitive to bitter tastes might eat high-calorie french fries in preference to lower-calorie broccoli.

One way geneticists look for patterns like this is through something called a genome-wide association study (GWAS), which is often used to identify genetic diseases. Researchers compare the genomes of people with and without the disease—Alzheimer's disease, say, or a strong family history of cancer—and look for parts of the genome that differ in the two groups. The disease-related genes they're looking for must lurk somewhere within those regions of difference. In the case of obesity, a GWAS would compare the genomes of overweight people against their normal-weight peers. Sure enough, those studies do turn up regions of difference, suggesting that there must be genes that affect obesity. However, not a single one of those regions of difference includes any taste receptor or odor receptor genes. How we perceive the flavor world doesn't seem to matter at all in determining our risk of obesity.

And there's yet another reason for suspecting that flavor, by and large, doesn't have a lot to do with how much we eat. If our overeating was driven by delicious flavors, then people who lose their sense of flavor—especially smell—should lose interest in food, and they should have a hard time eating enough. This was the question that led me to Monell's boardroom to eat tasteless hamburgers with Gary Beauchamp. And to see what happens after long-term loss of flavor senses, I headed just a few blocks down the street to the University of Pennsylvania Medical Center and Richard Doty's clinic on smell and taste disorders to meet some of his patients.

Patricia Yager had never had a serious health problem. "I go to Antarctica for a living. They don't let you go there if you aren't healthy," says Yager, an oceanographer who studies climate change and the oceans. In January of 2014, though, Yager—a slender woman with a broad face, heavy-lidded eyes, and long hair streaked lightly with gray—noticed a persistent metallic taste in her mouth. As scientists do, she worked through the possibilities: acid reflux, menopause, diabetes. Nothing fit. Her doctor found a little fluid in her middle ear and suggested decongestants, but the metallic taste persisted.

Then one day she was cooking in her kitchen when her preteen son came rushing in saying there was a terrible smell in the house. It turned out that some cheese had bubbled over in the oven and was burning. "I didn't smell the smoke," Yager recalls. "I thought, oh my gosh, something serious is going on here!" An ear, nose, and throat specialist guessed that her sense of smell was damaged, and the likeliest causes were permanent nerve damage or a brain tumor—not what she wanted to hear. "So I ended up in a puddle on the floor." Fortunately, an MRI ruled out the latter, scarier option, which is how Yager ended up in Richard Doty's clinic in Philadelphia.

Doty, a neuropsychologist, directs the Smell and Taste Center at the University of Pennsylvania, which is widely regarded as the best place in North America for diagnosing and treating smell and taste disorders. "We're a unique center in the world, really," says Doty. By the time patients make it to his center, many have already seen several doctors without understanding their condition, and they're desperate for Doty's specialized knowledge—even though he often can't fix the problem. "Much of what we do is correct misinformation and put people at ease," says Doty. "One of the things I like about this job is that most people are thankful that they came to see us, since we understand their problem."

A few days each month, he sees patients in his small, crowded office at the center. It's a classic academic office: books, papers, and binders cover his desk and side tables in teetering stacks a foot and a half high. The desk alone has six of these piles, and he and Yager have to peer between the towers to see each other. At first, Yager tells him, things either had no smell at all, or else they all had the same, unpleasant smell. Lately, though, she's noticed that she can sometimes distinguish among different smells. "None of them smell good yet, and none of them smell like they used to smell. Watermelon doesn't smell like watermelon, but it has a very distinctive, unpleasant smell." Vanilla now has an odd, turpentiney smell.

Most likely, Doty tells her, a viral infection has killed some of the nerve cells that carry odor receptors. We start out with a few million of these cells, each one carrying just one of our four hundred or so odor receptor molecules. A severe viral infection of the nasal passages can sometimes kill enough cells that some—or, in extreme cases, all—of the odor receptors effectively go extinct in the nose. The effect is like progressively cutting some strings on a piano: the chord begins to sound dissonant, then unrecognizable. Cut enough strings and the piano falls silent. It's also possible that Yager's smell loss is the result of a head injury, since she did hit her head in a fall while roller-skating a few weeks before she noticed the loss. That fall—mild though it seemed at the time—could have severed the connection between the olfactory epithelium and the brain.

Doty sends Yager off for a battery of tests to measure her senses of smell and taste. His associates measure the shape of her nasal cavity and the airflow through it; they test her sense of taste by dripping sweet, salty, sour, or bitter test solutions on each quadrant of her tongue, and by stimulating the tongue with an electric probe; they give her the University of Pennsylvania Smell Identification Test—the same scratch-and-sniff test that Doty had given me a few months earlier in Florida. The test is multiple choice, which avoids the complication of recognizing a smell but being unable to name it. That also makes it easier for Doty to spot malingerers who are faking smell loss in the hope of profiting from a juicy lawsuit.

While Yager is off doing the tests, Doty explains that viral infections are one of the three most common causes of smell and taste problems, along with head trauma and chronic nasal and sinus inflammation. Many of the patients he sees arrive in his office complaining of a loss of taste, but most of the time his tests reveal that the problem is actually with their sense of smell—further proof that most people can't really distinguish between their two main flavor senses. Olfactory defects are surprisingly common, affecting one in five people by most estimates, and about one person in twenty has no sense of smell at all. Often, people are completely unaware that they have a problem—in fact, one study found that asking people if they have a defective sense of smell tells you nothing useful about whether they actually do. (Oddly, older people tend to have lost olfactory ability without knowing it, whereas younger people tend to underestimate their sense of smell.) "Every time you get a bad cold, or are exposed to pollution, it takes a toll on the olfactory epithelium," he says.

Sometimes—as, apparently, with Yager—a single infection is enough to push a person "over the waterfall," as Doty puts it. Other times, the damage accumulates gradually, and our ability to smell slips away bit by bit, without our noticing, as we age. (Taste may fade a little bit with age, as well, but not enough for most people to notice.) Most of us, if we live long enough, will eventually have trouble with smell: Almost 30 percent of seventy-year-olds and about 60 percent of people over the age of eighty have significant impairments of their sense of smell, with men more likely to lose function than women. Surprisingly, scientists haven't tracked enough people through their lifetimes to be sure whether age-related loss creeps up slowly, or whether we reach a threshold where problems suddenly become much more common. Often, studies simply compare a group of older people with younger ones and report that the older ones have a worse sense of smell.

One sparkling exception to that rather dismal track record came in 1986, when every one of *National Geographic* magazine's 10.5 million subscribers received a scratch-and-sniff smell survey with their September issue. For each of the six odors, the subscribers were asked to rate the odor's intensity and pleasantness, and pick the best description of the odor from a list of twelve possibilities. They also answered some questions about themselves, so that the masterminds behind the survey—Monell researchers Charles Wysocki and Avery Gilbert—could make sense of their responses.

The survey was a huge success, with more than 1.2 million readers returning their questionnaire. When Wysocki and Gilbert tabulated the results, they found—as expected—that more older people than younger ones had trouble detecting some or all of the odors. Surprisingly, though, people's sense of smell didn't fade uniformly—they lost some odors faster than others. Virtually everyone could smell the banana, clove, and rose odors right up into their sixties, and even after that age, the ability to smell those odors trailed off slowly. Even among ninety-year-olds, 90 percent of men and almost 95 percent of women could still smell the clove and rose odors, and the success rate for banana was only a few percent lower. In contrast, the ability to smell mercaptans—the stinky chemicals added to natural gas to make people aware of leaks—began to drop off when people were in their forties.

Scientists aren't sure exactly why our sense of smell often fades as we age, but most of them think it's just part and parcel of our body's diminished ability to repair itself. The cells of the olfactory epithelium are among the few nerve cells that the body regularly replaces during adult life. As with other regularly replaced cells—skin and hair follicles are obvious examples—problems accumulate over time. The olfactory epithelium of a newborn infant is a nice smooth, solid sheet of cells, but it gets more ragged and patchy as we age.

But something else may be going on, too. Even as the olfactory epithelium breaks down, the responses of its remaining cells may start to blur. To show this, a team led by Beverly Cowart, yet another Monell researcher, collected biopsy samples from the olfactory epithelium of elderly and middle-age volunteers. This is as creepy a procedure as it sounds. Under local anesthetic, doctors thread a fiber-optic scope into one nostril and insert something called a "Kuhn-Bolger giraffe forceps"—a scissorslike clamp with a long, offset neck—up the other nostril to grab a little pinch of olfactory epithelium. The resulting cells can be grown in petri dishes to see which odors—or, in Cowart's case, odor mixtures—they respond to. Each cell from middle-aged noses responded to just one of the two odor mixtures Cowart used. By contrast, about a quarter of the cells from elderly noses responded to both. That suggests that older people's noses blur together details they once could have resolved—an olfactory analogue of cataracts, perhaps.

The healthier we stay, though, the more likely we are to keep our sense of smell intact: The "successfully aged elderly" often continue just fine. Smell loss, in fact, can be an early warning sign of more serious medical problems such as Alzheimer's disease and Parkinson's disease. That's not surprising: the olfactory system is basically part of the brain, so many degenerative brain diseases should affect the sense of smell, as well. Oddly, some oncologists also report that one of the first warnings of a developing cancer is that food doesn't taste right—even when the tumor is in the breast or prostate or some other organ unrelated to flavor perception. Indeed, seniors who have lost their sense of smell are four times as likely to die within the next five years, compared with people of the same age with good olfaction. (It's important to note that olfactory loss is not a death sentence—most people who lose their sense of smell still survive just fine.)

Whatever the cause, loss of smell can bring major problems.

In one study, nearly half of the patients with smell disorders reported experiencing depression and anxiety, and more than half felt isolated and had difficulties relating to other people. The effect on flavor is even worse, with 92 percent of people reporting less pleasure in eating—and that brings social difficulties of its own. "Most of our social interactions involve food," says Cowart. "It becomes very difficult to justify going out to eat and paying a lot of money for food they can't taste, or going to a friend's house and not being able to tell the host that the meal tasted great."

You could hear that loss in the voices of the patients in Doty's office—though, being nonspecialists, they tended to talk about failures of taste, not smell. "I don't taste my food at all," said one elegantly dressed older woman. "It tastes like eating sawdust when I eat a cracker." Or, as another put it, "The only reason I know what I ate is that I'm looking at it, and I remember what it tasted like."

Despite complaints like this, most people find ways to cope, and about two-thirds manage to maintain their usual weight. Only a small minority—one expert puts it around 10 percent—of people with a damaged sense of smell actually lose weight as a result. And those tend to be people who suffer not total loss, but distortion of smell, like Yager's turpentiney vanilla and stinky watermelon. Often, patients report that everything has the same "burnt-chemical" smell—probably the best they can do at describing something unfamiliar and unpleasant. Olfactory bugaboos like this can actively turn people off their food. Among the elderly, people who have lost their ability to smell are far more likely to be undernourished—but that may be because loss of smell is linked to other health problems, rather than because they find their food less appealing.

On the other hand, a few people who suffer a sudden loss of smell actually gain weight as a result. These tend to be people who were already susceptible to food cravings, which are often more about habit than about deliciousness. (One researcher who tried to create cravings for other foods by putting people on a boring diet of vanilla-flavored meal-replacement drinks found that the subjects actually began to crave the boring drinks. "They tried to scam cans of this stuff off the technicians in the lab," she recalls.) The craving gives them a sensory template, a hole waiting to be filled—and with no sense of smell, they keep eating in the futile hope of finding satisfaction.

When Yager returns to Doty's office that afternoon, he gives her the results of the tests: no problems at all in her sense of taste, but on her smell tests she scored no better than if she'd been guessing randomly. Clearly, what little smell she has left is so badly degraded that she can't tell the difference between familiar odors like grape and peanut butter.

And unfortunately, says Doty, medical science can't do much to fix the problem. About half of the people who suffer smell impairment get some function back within a few years, but less than a quarter recover fully. Among people with complete loss of smell, like Yager, the odds of full recovery drop to just 8 percent.

There might be ways to improve that discouraging prognosis, he notes. There are a few reports that a supplement, alpha-lipoic acid, might help. And some studies suggest that even a failing sense of smell might improve with practice, because nerve cells are more likely to be replaced or regrow if they're being used. Grab bottles of spices—"Anything that says McCormick"—and keep them beside your bed, he tells Yager. Sniff through them three or four times first thing in the morning and before you go to bed for the next three or four months, and see if it helps. She brightens at the thought that she might be able to do something about the problem.

A year later, I checked in with Yager to see whether the exercises helped. No luck, she reported—she still can't smell anything. "I've grown accustomed to it, I suppose," she says. She's learning to cope by making sure her food is well seasoned with salt, pepper, and lemon, which don't depend on smell for their flavor impact, and admits that "Sriracha has become a close friend." (The chili burn uses a different nerve to reach the brain, so she still gets full value for that flavor component.) She rarely drinks wine any more, except to be sociable, since it doesn't offer much to interest her. These days, her preferred tipple is gin and tonic, which is still exciting in her mouth, thanks to its pronounced hit of bitterness.

It's hard to come away from Doty's clinic with the feeling that flavor makes much difference in maintaining body weight. But what does, then? Why do some people gain weight while others don't? And, the big question: What can we do about it? Unfortunately, scientists still haven't agreed on an answer. Mark Friedman thinks—and he has some research to back him up—that overweight people's metabolism has shifted so that the energy they take in from meals is more likely to get stored as fat, and less likely to be available for the daily needs of living and moving around. "You're losing energy internally that you can't get to, so you eat more," he explains. "Essentially, you're overeating because you're getting fat."

On the other hand, Dana Small thinks—and there's a lot of evidence to back her up—that overweight people are less sensitive to their body's satiety signals, so that they're less likely to shut down their food intake when they should. Instead, they tend to eat from habit, because it's time, or they walk through the kitchen, or because they drive near the golden arches. In the absence of a good satiety signal, they're more vulnerable to the pull of their reward system, even when they're not actually hungry. Even rats sometimes overeat merely because the food is there. In one study, just putting extra containers of sugary or fatty food in a rat's cage was enough to get it to eat more of the abundant calories. Similarly, rats that can choose among six different water bottles get almost twice as fat if five of the bottles contain sugar water than they do when just one bottle has sugar water. They could have drunk just as much sugar water from the single bottle—the researchers kept it full—but something about having so many sugary options made a difference.

The bottom line here seems to be that flavor does make a difference to what we eat, and indirectly to how much we eat. Flavor-nutrient conditioning pulls us toward wanting the calorie-laden foods that are so readily available these days. But even though flavor is part of the overeating equation, tinkering with flavor may not be part of the solution. Making food more or less flavorful probably won't change consumption in the long run. No matter how enticingly you flavor that no-fat rice cake, your body will still know there's nothing much there—and will quickly learn that those flavors aren't worth liking. Given a choice between a full-fat cheese and a reduced-fat version, the reward systems in our brains will clamor for the full-fat one. Any food company that tries to sell the reduced-fat version has to fight this innate emotional pull—usually by appealing to reason and prudence, which have a hard time competing against yearning. Often, food companies don't even try, and when they do they frequently lose. That's why frozen pizzas have as much cheese as they do, and why French fries remain so popular on fast-food menus. They're the intensely researched result of the flavor companies that design, make, and test the flavors in processed foods.

Chapter 6

WHY NOT IGUANA?

When Bob Sobel's kids were little, they would sometimes run up to a strange man in the grocery store and hug his leg, saying, "Hi, Grandpa!" Eventually, Sobel figured out the reason for the error: Their mental sketch of "Grandpa" consisted of gray hair, glasses, and a beard—and anyone who satisfied those criteria, they figured, must be Grandpa. It didn't take many failures before Sobel's kids upgraded their sketch, of course, but he has always remembered how little information they needed at first to leap to their conclusion.

It's a lesson Sobel—no relation to Noam Sobel of the chocolate-soaked string—puts to use every day in his job, as vice president for research at FONA International, a company in the business of creating flavors for the food industry. Designing flavors is largely a matter of finding a way to sketch a chemical likeness—a caricature, if you will—of reality. Case in point: Sobel likes to give people a fresh apple and a Jolly Rancher green-apple-flavored hard candy. "Which one has more chemicals?" he asks. Most people assume that it's the patently artificial Jolly Rancher. But nature is made of chemicals, too. The real apple, in fact, contains at least twenty-five hundred different flavor chemicals, while the Jolly Rancher has precisely twenty-six. What makes the flavor industry possible is that our mental image of "apple flavor" doesn't require all twenty-five hundred chemicals. "Just like our picture of Grandpa, it's going to pick out a few," says Sobel. That's exactly what Jolly Rancher has done with their apple candy. "It has enough information to give you the apple. The goal of flavor chemists is not to duplicate exactly all 2,500 flavor chemicals that nature uses. It's to re-create the impression."

Sobel is explaining all of this in his well-modulated, gently sibilant voice in an auditorium at FONA's headquarters in Geneva, Illinois, a bucolic suburb of Chicago. In an industry notorious for its obsessive secrecy, FONA is unique in flinging wide its doors to let the sun shine in. Several times a year they welcome clients, competitors, and the odd hanger-on like me to Flavor 101, a free short course in the workings of the flavor industry.

In the room with me are two chewing-gum developers from Wrigley, people from Butter Buds Food Ingredients (a manufacturer of dairy flavorings), Grapette (a maker of soft drink flavors), and PepsiCo (which needs no introduction). There's someone from a company that makes vegetarian "meat" products. There are representatives of processed-food makers, pharmaceutical companies, a major liquor company, and a food-packaging company. There are several new hires at FONA itself. And, along with me, one other outsider—an anthropologist studying the food industry.

Sobel's long face, slightly protruding lower lip, and pleasant smile make him look a bit like a 1980s-vintage TV news host. He has the enthusiastic demeanor of a good high school chemistry teacher, which is what he once was. Back in 1999, his wife suggested he take an outside job during the summer holiday. He ended up working as a flavor analyst at FONA and discovered a world he'd never known of. Entranced, he's been there ever since.

That's a sentiment I encountered over and over from professional flavorists. Playing with chemicals to concoct a flavor is a bit magical and a lot of just plain fun. It's applied chemistry of the most appealing sort. Most chemists work with unpleasant, often toxic substances and go to great lengths to avoid inhaling or ingesting their products. Flavor chemists, on the other hand, do it all the time.

Flavor chemistry is also a big, big business. Flavor companies sell more than $10 billion worth of flavorings every year, and the products that result can be found in nearly every kitchen. Almost every convenience food, almost every processed food, almost every fast-food product relies on added flavorings, both to make the food more appealing and to provide consistency from batch to batch. Flavorings are the reason your favorite bottled spaghetti sauce always tastes the same, even though one batch of tomatoes might be sweeter and more fragrant than the next. Flavorings help your strawberry yogurt taste like strawberry yogurt rather than just yogurt with strawberries in it. Flavorings help diet-food companies keep their products appealing even as they reduce the fat. And, according to some critics, flavorings interfere with our bodies' natural ability to select a balanced diet, which could make them a key player in the modern epidemic of obesity—a charge we'll return to later.

The modern flavor industry really began in the 1950s, when chemists developed a tool that let them separate, sort, and identify the individual molecules that make up a flavor. This tool, called a gas chromatograph, separates the molecules by how fast they travel through a long, coiled tube—a speed that depends on the molecule's size, shape, and electric charge. If the tube is long enough, each kind of molecule will finish at a different time, and a chemist waiting at the end can catch and identify them one by one as they emerge.

Suddenly, flavor chemists had the detailed knowledge they needed to take flavors apart and build them up again, brick by brick, instead of relying on crude extracts of natural products. The design of flavor changed from an arcane art into a quantitative science. As chemists built up their understanding of which molecules contribute which aromas to a flavor—methyl anthranilate smells like grape, gamma-nonalactone like coconut, furfuryl mercaptan like freshly ground coffee—the flavorists' tool kit exploded in size. Today, a well-equipped industrial flavorist can choose among over seven thousand different molecules and extracts when assembling the components of a flavor.

Learning how they do that is why I'm here in Flavor 101, listening to flavorist Menzie Clarke explain how she puts together a flavor. Clarke is a small woman of Asian ancestry with a broad smile and boundless enthusiasm for her craft. This enthusiasm surfaces in her rapid speech, as her words tumble out half formed, propelled by the pressure of her racing thoughts.

For many flavors, she says, you start with a so-called character compound—a molecule that shrieks out a particular flavor so loudly that it's almost impossible to build that flavor without it. If you smell amyl acetate, for example, you'll instantly recognize it as banana. Likewise eugenol and clove, or citral and lemon. If your flavor has a character compound, you're halfway home already.

Next, you layer on some "top notes," the up-front quick hits of flavor that burst onto the palate but fade quickly. These lack the immediate recognition of the character compounds, but often deliver a more generic quality. For example, ethyl butyrate delivers a fresh, fruity top note to many citrus flavors. Bottom notes, in contrast, build slowly but linger longer—vanillin is a good example, or the creaminess of delta-lactones—to add fullness to the flavor.

With the skeleton of the flavor in place, you start to think about differentiators, the elements that add subtle highlights to the particular flavor you're building. If you want a slightly mealy note to an apple flavor, for example, you might add a little tagette oil. If you'd prefer a greener note, use a bit of cis-3-hexenol instead. Add a little furaneol for more of a baked-apple flavor—or add a lot for a candy-apple flavor.

Finally, you pay attention to the balance of the flavor. "You don't want your flavors to have spikes," says Clarke. "You want them nice and balanced, a very clean flavor." Often, that means keeping things simple—but Clarke's "simple" seems an awful lot like my "complex." "You don't want to use more than 30 to 40 flavor components," she says. "Once you get to more than 40, it gets kind of messy. You wonder if it's really necessary."

The process sounds straightforward, but of course the reality is often a lot more complex. Sometimes, for example, key flavor molecules turn out to be very short-lived. A molecule important in the flavor of fresh watermelon, for example, breaks down within thirty seconds of release, so it can't be used in a commercial watermelon flavor. "Everybody wants to have that fresh watermelon flavor," says Sobel. "The problem is, you only get that flavor when you actually bite into a watermelon."

That's not the only example, either. Short-lived 2-acetylpyrazine provides the ephemeral popcorny aroma in freshly made basmati rice, which flavorists cannot reproduce successfully. The furfuryl mercaptan that makes the character note of freshly ground coffee also vanishes quickly. That's why the first whiff of a freshly opened can of coffee is so much better than reopening the same can the next morning. (It's also why coffee tastes so good in a coffee shop—all the grinding and brewing they do ensures a steady infusion of furfuryl mercaptan into the air, where it can enhance your flavor experience.)

The next morning, in Sobel's office, I ask Clarke if she would talk me through a real flavor formula. I'm not hopeful, because most formulas are closely guarded trade secrets, but Sobel surprises me: He's able to pull out an example that's publicly available, a pineapple flavor originally developed by International Flavors and Fragrances, one of the big flavor companies. It's not too complex, having just sixteen ingredients, so it seems like a good choice for analysis.

Clarke recognizes the flavor as pineapple—even before anyone says the word—because of the presence of allyl caproate, a character compound for pineapple. "If I see allyl caproate in a flavor, I directly go into pineapple mode," she says. Then Clarke starts to pick apart the rest of the recipe. Ethyl butyrate and ethyl acetate ("the ethyls," she calls them) supply generically fruity top notes. A set of three acids—acetic, butyric, and caproic—also add bright top notes. Acetic acid, of course, is vinegar. Caproic acid smells a bit goaty, while the odor of butyric acid is often described as "baby vomit." (Perfumers know that a bit of something nasty—cat pee, for instance—can often add a little depth and complexity to a fragrance; the same is true in the world of flavor. Wine connoisseurs often discern a whiff of cat pee in the aroma of sauvignon blanc wines.)

Next come a couple of chemicals—terpinyl propionate and ethyl crotonate, if you care—that contribute a husky, rindy character to the flavor. These probably serve as differentiators, helping to make this particular pineapple flavor a little different from all the others.

The rest of the flavor formula consists of tiny amounts of several essential oils—oil of sweet birch, oil of spruce, oil of orange, oil of lime, oil of cognac, and others. Instead of being single chemicals, each of these oils is a mix of many different flavor compounds and, as their names imply, are usually extracted from a natural source. "These are to be creative," says Clarke. Differentiators again, in other words. Some, such as oil of cognac, also supply heavier, lingering bottom notes to the flavor.

But the ingredients list alone isn't enough to make a flavor. You also need them in the right proportions—and that can be tricky. Should the allyl caproate be 5 percent by weight, or 4 percent or 6 percent? You'll have to test it to know for sure. And there are other pitfalls, too. Merely doubling the concentration of a flavor molecule doesn't always double its intensity. Sometimes the quality of the flavor changes instead. Linalool often gives a nice blueberry character at a concentration of .02 percent, for example, but at .025 percent it can lose its blueberryness and take on an unbalanced floral quality—a phenomenon flavorists call "flavor burn." (This means that food companies can't just crank up the flavor dial to compensate for an aging population's fading senses. They'll need to rebalance every flavor at its new intensity—a much bigger task.)

Flavor 101 was a great way to learn the basics of the flavor industry—but to dig deeper into its intricacies, I needed to get my hands dirty. I headed east on a pilgrimage to the epicenter of flavor.

Cincinnati, Ohio, seems like an unlikely candidate for the flavor capital of North America. It's an unassuming midwestern city full of unassuming midwestern people of largely German extraction who live in unassuming midwestern two-story brick houses with friendly front porches and well-kept lawns. Gastronomically, its claims to fame are a pork-and-oatmeal breakfast sausage and something called "Cincinnati chili," which is not chili at all but a cinnamon-spiced meat sauce usually served over spaghetti or hot dogs. Yet just a short drive north from the center of town, you'll find a nondescript industrial park with several unassuming brick-and-glass buildings that house the U.S. headquarters of Givaudan, the world's largest manufacturer of flavors.

You've undoubtedly tasted some of their flavors. We all have, or at least everyone who's ever bought a food product other than a raw fruit, vegetable, or meat, or drunk anything other than water, beer, or wine. Givaudan's flavors show up in soups, soft drinks, cookies, candies, frozen dinners, fast food, and almost any other food product you can think of. Yet you'll never find their name on the label. Nor will anyone from Givaudan ever let slip the name of any product that uses their flavor. Dr Pepper/Snapple, the giant beverage conglomerate, has a factory right across the parking lot from Givaudan's facility. Givaudan's spokesman says it's entirely coincidental that a drink manufacturer happened to set up shop right next to a flavor developer, but he also can neither confirm nor deny that the company is one of Givaudan's clients. The level of secrecy involved would make the CIA proud.

I've been trying for more than a year to arrange a visit to one of the big four flavor companies, or "flavor houses," as they're known in the trade. (Besides Givaudan, the others are Firmenich, International Flavors and Fragrances, and Symrise. The flavor industry also includes a dozen or so middle-tier firms—FONA among them—and dozens of tiny flavor houses, often specializing in a niche market like grape or dairy flavors.) Mostly, it's been a long, frustrating sequence of unanswered e-mails, unreturned phone calls, and general silence. They just don't want people to know. Finally, though, I got lucky. Someone I met at a conference knew someone who had just retired from Givaudan, who must have pulled some strings with Jeff Peppet, the company's communications officer. Suddenly Peppet, who had been ignoring my e-mails and voice mails for months, actually got back to me and offered—O Fortuna!—to arrange a visit. And so, to my astonished disbelief, here I am at last, parking my car and walking in the front door of Givaudan.

In person, Peppet—who appears to be in his midforties, with expensively cut hair—couldn't be more helpful and welcoming. He's set up a full day's worth of interviews for me, spanning a large part of Givaudan's flavor development work. (He even warned me away from the Cincinnati chili when I asked for dinner recommendations.) But the part I was most interested in was a session with flavorist Brian Mullin, who was going to let me build a simple flavor for myself.

Mullin's about sixty, with a full head of graying hair; wide thin lips bracketed by deep smile lines; and a firm, friendly gaze. He's got the slightly raffish demeanor of a favorite uncle who's always enjoyable. He insists on shaking my hand, even after I mention I have a slight cold, instead of yanking his hand away hastily like every other flavorist I met. (A flavorist with a cold is like a warehouse worker with a bad back—they can't do their job, and have to fill the time with paperwork.) It's good to challenge your immune system, he says.

The first step in making a flavor, Mullin tells me, is to be clear about what the client wants. Suppose, for example, that I had come to Givaudan saying I want a strawberry flavor. Well, fine. They already have thousands of strawberry flavors. Do I want a ripe one, a green one, an especially fruity one? Do I want a simple, inexpensive recipe or a costly but more realistic version? The client's answers to those questions will help determine the right starting point. I think of the best strawberries I've ever eaten, the ones I used to buy at the farmer's market near where I lived in coastal California. The strawberry fields were just down the road, and I'm convinced that the berries that were picked too perfectly ripe to ship ended up in our market. Their fragrance was powerful enough to seduce you from clear across the parking lot. That's the strawberry I want.

No matter, though. Mullin's already picked the recipe we'll use for our demo. He hands me a single sheet of paper with a short list of ingredients. "Mother Nature's already decided what goes into a strawberry," he tells me. Of course, no customer could afford to include all the hundreds of flavor compounds found in a real strawberry, nor would they need to. The trick is to pick the key ingredients that will bring our flavor close enough to the real thing, for a price we can live with. For many flavors, you'd start with the character compound: amyl acetate for banana, methyl benzoate for cherry, citral for lemon. But strawberry has no character compound—there is no single molecule that smells like it, so even the simplest strawberry flavor has to be built up from several components, each of which contributes one facet of what we'll perceive as strawberry. Mullin's recipe has just four ingredients—simple enough for me to make quickly, but sufficient to clearly sketch out the flavor of strawberry.

Before we go into the lab, Mullin introduces the ingredients to me one by one in his office. The first is ethyl butyrate. He grabs a brown glass bottle off his desk, unscrews the cap, and dips a strip of filter paper into the liquid inside. Then he offers the strip—flavorists call it a blotter—to me to sniff. It has a bright, generically fruity aroma, and it supplies the essential top note of our flavor.

Flavorists sniff a lot of blotters in the course of their work, enough that most of them carry books of blotters in their pockets like smokers carry matches. (Mullin's still bear the logo of a previous employer, a flavor house he last worked at seven years ago.) Like almost every flavorist who ever passed me a blotter, he warns me not to let it touch my nose as I sniff—a drop of concentrated odorant on the nose is just as disabling to a flavorist as a sprained ankle is to an athlete. There's also the question of what to do with dipped blotters after smelling, especially if you might want to sniff them again in a few moments. Most flavorists I visited just set the blotters on the corner of the desk, which risks a lot of aromatic contamination of the desk surface. Mullin, however, uses an old hand's trick of the trade: he creases the blotter with his thumbnail just below the dipped part, so that when I set the blotter down, its moist tip sticks safely up from the desk.

Item two in the recipe is cis-3-hexenol. Mullin dips another blotter and passes it to me. This one smells exactly like freshly mown grass, and adds a green note to the flavor. (Look for that grassy green note the next time you eat a strawberry. You may not have noticed it before, but it's there.)

Next up is furaneol, which supplies a brown, cotton-candylike sweet smell characteristic of ripe strawberries. "If you make a strawberry without furaneol, in my opinion, you'll never sell it," says Mullin. "The more you put in, the better it is—but there comes a point where you can't afford it." Furaneol gives the lingering finish that helps define a good strawberry flavor. "It just carries and carries and carries," he says.

The fourth and final ingredient in our flavor is gamma-decalactone. On the blotter, it smells a bit peachy. It's there, Mullin says, to fill a gap in timing: ethyl butyrate hits right up front, followed quickly by cis-3-hexenol, but furaneol's contribution takes a while to develop. That could leave a hole in the flavor, which the gamma-decalactone fills.

I now have four blotters on the desk in front of me, tips raised off the surface like a family of baby cobras. Following Mullin's instructions, I gather up all four and waft the bundle under my nose. Presto: strawberry! Not the strawberry of my dreams, the farmer's market berries from California, but a recognizable strawberry, nonetheless—and further proof, if any is needed, that a skilled flavorist can assemble a flavor that smells nothing like any of its individual components.

Root beer is another good example of this. Once upon a time, as you might guess from its name, root beer was made from an extract of sassafras root. But safrole, the main aromatic oil from sassafras root, turns out to be carcinogenic, and in 1960, the United States banned its use in soft drinks. Root beer manufacturers had to concoct their flavor some other way, and Mullin shows me one option: a top note of methyl salicylate (which smells like wintergreen Life Savers), a middle note of anise-smelling anethol, and a lingering base note of vanillin. Put them together, and it's unmistakably root beer. It surprised me to learn that root beer's top note is wintergreen. I'd never noticed that before, and I doubt many others outside the flavor industry have, either. But once you know to look for it, it's definitely there. (Actually, most Europeans—who didn't grow up drinking root beer, and thus don't instantly recognize that blend of flavors—get it right away. Many of them can't imagine why we drink the stuff in North America, because to them it smells like the wintergreen-scented liniment you rub on sore muscles. "Why would you want to drink something that smells like a rugby locker room?" one Brit asked Bob Sobel on encountering root beer for the first time.)

But enough sniffing. It's time to hit the lab. Mullin grabs a lab coat off a hook behind his door and gives it to me, along with safety glasses and a handful of plastic eyedroppers. "Let's go make a strawberry," he says. We're making a test-sized batch, the amount a flavorist might produce while tweaking a formula, so the process turns out to be a simple matter of measuring fluids into a beaker. Just a tiny bit of ethyl butyrate and cis-3-hexenol, 0.08 grams of each, which turns out to be somewhere between three and four drops. Mullin suggests measuring those first, so that if I accidentally put too big a squirt into the beaker, I won't have to discard a large volume of ingredients. Then fifteen grams—about a tablespoon—of urine-yellow furaneol and a squirt of gamma-decalactone. Stir, then dilute with water.

Now it's time to see what we've made. You'll recall that we sometimes perceive retronasal flavors rather differently from orthonasal smells. As a result, serious testing of a flavor formula almost always happens by actually drinking the concoction—no mere sniffing at this stage. Sipping the flavor, I find it a little disappointing. It doesn't have the ripe, ripe oomph I was hoping for, and the green note—inconspicuous on sniffing—comes through too strongly in the mouth. The next step, says Mullin, would be to modify the recipe by using a little less cis-3-hexenol and a little more ethyl butyrate next time, to see if that gets closer to my target.

In practice, this trial and error would go on again and again, with repeated taste testing until the client is finally happy with the result. It's a slow process: Mullin's assistant can mix up perhaps a dozen formulas in a day, particularly if they're more complex, so settling on a final flavor could take days or weeks. That makes a custom flavor like this an expensive proposition.

To speed things along, Givaudan has developed ways to automate the process somewhat. Earlier that day, another of their researchers, Andy Daniher, had showed me a suitcase-sized device they call the MiniVAS (for Virtual Aroma Synthesizer) that Givaudan flavorists can take out on house calls. The device has slots for thirty vials containing aroma "keys," which can be single odorants or complex mixes such as lemon-peel extract or cola flavoring. By moving sliders on a touch screen, users can change the proportions of each key in the mix, then see how the aroma changes. (The MiniVAS has three output ports shaped like the negative of a nose, so that flavorist and clients can all sniff at the same time.)

"Let's talk spiced rum," says Daniher. With a touch of his finger, he starts air bubbling through a rum base. Another touch adds a hint of strawberry to the rum—a terrible idea, we all agree. Two more quick touches and the strawberry is replaced by orange. Much better. "Now you can start to say 'I like this, I don't like that.' You can just show what you like about a flavor," says Daniher. "You can do a lot of flavor-creation work very quickly and zero in to a formula you can compound." Best of all, the whole thing can be controlled remotely, so that a flavorist here in Cincinnati can collaborate with another flavorist in Asia and a client in London, and all can smell the same thing simultaneously.

For many customers at the large flavor houses, all this analysis may be overkill. They may be able to bypass flavor development entirely—and save a lot of money in the process—by using an off-the-shelf flavor. At Givaudan, these clients end up talking to Laurence Roquet, who manages what the company calls its "portfolio"—a searchable library of previously created flavors. Roquet is French, and she looks it—tall and slender, with bobbed, black hair and a round face emphasized by her huge, round glasses. "Why re-creating the wheels?" she asks, in her fluent but slightly askew English. "Why make another strawberry when you have hundreds of strawberries on the shelves? We have so many flavors, good stuff inside. Why not do that?"

Givaudan's full portfolio might have one hundred thousand flavors in it, though the core portfolio that they use regularly is around three thousand flavors. Each one of the flavors in that core library is assigned a series of tags that describe the flavor itself (juicy, pithy, and so forth), its possible applications (sweet, savory, cold beverages), and its regulatory status (is it organic, natural, GMO-free, approved for alcoholic beverages). That lets Roquet and her staff quickly pull up a short list of flavors that meet a client's requirements. Then it's time for tasting. Often, the client finds a flavor that they're happy with as is. If not, the portfolio at least suggests the right starting point for further tweaking. About 70–80 percent of Givaudan's flavor projects start at Roquet's desk, she says.

Pulling ready-made flavors off the shelf is one end of what you can think of as the innovation spectrum. At the other end, Givaudan also puts a lot of effort—and plenty of cash—into discovering and re-creating new flavors. Often, this involves prospecting in the natural world, looking for fruits, flowers, or other plant parts that offer new flavor molecules to beguile the world's palates. One favorite source is the botanic garden at the University of California, Riverside, which hosts the world's largest collection of citrus trees. By sampling fruits in the Riverside collection, Givaudan's flavorists have found several new citrus flavors, including a sweet lime with just a hint of pepper. "That was nature showing us something we would not have thought about," says Daniher. Finding these gaps—white spaces on the flavor map, even in a region as well traveled as citrus—is always amazing, he says.

Sometimes Givaudan's explorers go further afield. Years ago, Peppet participated in an expedition to Gabon, Africa, where Givaudan chartered a blimp to float above the rain forest canopy so that technicians could collect scents from every flower and fruit they could find. Back home, flavorists sorted through it all, looking for elements that they could add to their arsenal of flavor chemicals.

Other times, they don't have to go far at all. "It occurred to us that we don't have to go into jungles," says Daniher. "We can go into restaurants." Givaudan technicians order up an interesting dish, something with a flavor that they're interested in replicating. This is what they call their "gold standard"—the real thing, the target flavor they will try to approximate in the lab. The whole order goes into a chamber that captures the aromas rising off the food. Then the technicians analyze this "headspace" to figure out what makes it tick, so that flavorists can find ways to re-create it in their lab.

Daniher opens a vial labeled "kalbi flavor" and hands it to me. It smells just like the grilled meat, redolent with soy sauce and garlic, that's so delicious in a good Korean restaurant. "What I like about this is you can really smell the fatty, grilled notes," he says. But this wasn't extracted from real Korean barbecue—Givaudan's flavorists have re-created it from individual chemical components to match the headspace analysis. It's basically the gold standard in a bottle—an almost perfect match, but probably too expensive to be practical as a commercial flavor. Now it's up to their flavorists to develop a cheaper version that delivers nearly the same effect.

Another project they're working on is a flavor element that Daniher calls "richness." "Richness is what you get from slowly cooked foods," he explains. "We've all tasted a great stew that's been cooking for a long time." Daniher's researchers think they now understand which flavor molecules are responsible for this long-cooked flavor. When pressed for more details, Daniher clams up. "Proprietary stuff," he says. In essence, Givaudan may have isolated the flavors of time, care, and patience. If they're right, they could be on to something really big.

A job like re-creating Daniher's *kalbi* flavor or "richness" is likely to end up on the desk of someone such as Mary Maier at Givaudan. The world of industrial flavor is so broad that flavorists tend to carve out particular corners to specialize in. One person I spoke with had spent a long, illustrious career specializing in sweet brown flavors: honey, maple syrup, cola, and the like. There are fruit flavorists and beverage flavorists, dairy flavorists and candy flavorists. One of the biggest divides tends to be between sweet flavorists and savory ones. Maier, a senior flavorist, is one of the latter. Working in a world of meat flavors is tougher than doing fruit flavors, she says, because the flavors themselves are more complex. "There's not just one molecule that you smell and say 'Aha!'" she says. Maier is a short, fit woman with straight brown hair down to her shoulder blades, held back by a thin headband. Remarkably, she's a second-generation flavorist—as a college student, she used to mix up samples for her father, who also worked for Givaudan, and she ended up making a career there herself.

Much of Maier's work involves riding herd on the Maillard reaction, a complex network of chemical changes that happens during the browning of proteins and sugars. But where you and I typically start our Maillard reactions with a piece of beef or chicken, professional flavorists like Maier often start with protein extracts such as autolyzed yeast extract, or even pure amino acids and sugars, to give better control of their results. Start with the amino acid cysteine and the Maillard reaction proceeds toward chickeny-meaty flavors. Start with methionine and you get something potatoey-cabbagey; phenylalanine gives honeylike flavors or—in combination with the sugar fructose—a flavor that some describe as "dirty dog." (There's that paradox again: a little bit of something obnoxious adds interest to a complex flavor.)

We head out into Maier's lab to taste some flavors. First up is a chicken flavor that one of her customers wants to put into a powdered soup mix. Maier spoons some of the mix into a beaker, adds water, heats it on a hot plate, and offers me a spoonful. I taste onion, celery, and some doughy or grainy stuff from the noodles, but Maier says that's all irrelevant to the task at hand. Those flavors are part of the customer's base soup. Her job is the chicken flavor; from her perspective right now, everything else is just noise. Concentrating now on the chicken, I think I get some roasted-chicken notes, but Maier corrects me. What I'm tasting are the astringent, bony, fatty notes of a boiled chicken, not the caramelized brown, sulfury notes of a roasted one. The flavor is not bad, though: she's getting close to the target she wants.

Next project in her lab is a chicken patty, which will be breaded, prefried, and frozen, then baked at home by the consumer. There's a version of these already on the market, but the manufacturer wants to change ingredients. Maier doesn't know why they want to do that—to save money? To use more readily available ingredients? It doesn't matter. Her mission is to make the new version taste the same as the old one.

So her technicians have whipped up a test batch. There's a blank—the unflavored chicken nugget, which tastes generically chickeny. There's the original version, the target. And there's the test version, with the current version of Maier's new flavoring. On tasting the test patty, one of the technicians immediately says "That's extremely strong!" This is the first time they've tasted the new flavor on chicken, and it wasn't as aggressive during earlier taste tests in water—further proof, if any were needed, that there's no substitute for tasting a flavor in its final setting. Flavorists can't do their jobs in the abstract.

Maier takes a bite of the target next, and pauses to consider. "I get a precursor in there," she says. That is, she tastes one of the ingredients in the Maillard reaction that hasn't fully reacted—a sign that the client's original Maillard starting point wasn't quite right. She doesn't say it aloud, but I can imagine her thinking that whoever designed the original flavor did a sloppy job.

After some discussion, the team agrees that the target also has a more grilled, sulfury-meaty character, while the test nugget is more smoky and livery. The lack of grilled flavor in the test version disappoints Maier. "This is just softer. We're not getting that impact," she says. They decide to do another version next week to see if they can get closer to the target. Meanwhile, they'll also send the target off for analysis, to try to identify the precursor that's out of whack.

As Maier's chicken nuggets demonstrate, building a flavor that is balanced and convincing when tasted by itself is only half the job. The context a flavor is used in—that is, the other ingredients in the product, known in the food industry as the "base"—makes a huge difference to the final result. Many fruit flavors, for example, stand out more prominently in a sweet base, because we expect fruity and sweet to go together, and the brain amplifies these congruent stimuli. Similarly, a salty base brings out the savory elements in something like a chicken soup.

Another issue is that many flavors interact physically or chemically with the base. Even something as simple as a thickening agent, for example, can slow down the release of flavor molecules in your mouth, so that a thick drink or sauce would taste blander than a thin one with the same amount of added flavor. A lot of flavor molecules dissolve more easily in fats than in water, so a high-fat food also releases its flavor more slowly and may therefore need a higher dose of flavoring to achieve the same effect. At FONA, Bob Sobel likes to demonstrate this by mixing up identical amounts of instant chocolate drink in four different kinds of milk, ranging from skim milk to half-and-half. The differences are striking. Chocolate made with skim milk gives an intense burst of chocolate flavor that vanishes almost instantly. "It comes rushing out—it's not balanced," says Sobel. With 2 percent milk, the initial hit is less intense, but the flavor lingers a bit, and that's even more true of whole milk. The chocolate made from half-and-half, in contrast, has a much more muted flavor, but its richness lasts and lasts. Which is best? Try the experiment yourself at home, and see which you prefer.

Even after flavorists have built and balanced their flavor perfectly for its chosen base, the job's not done. There's one more big problem to solve: delivery. Sometimes you can't just dump the finished flavor straight into the food—adding liquid flavor to, say, instant oatmeal would result in a gummy mess. And often, the flavor needs to be protected so that it survives the journey from manufacturer to mouth. Exposure to air can oxidize some flavor molecules. Others—especially the volatile top notes of a flavor—are prone to just drift away, so that the flavor loses its oomph over time. Flavor decay can also happen in protein-rich foods because the sulfur atoms within proteins gradually latch on to the flavor molecules and prevent their release in your mouth. (This binding by proteins is why the smell of campfire smoke can lurk on your [protein-rich] hair, emerging when your hot shower adds enough energy to knock some scent molecules loose again.) And occasionally, the flavor and the food simply declare war on each other, such as when garlic oil prevents bread dough from rising.

The answer to almost all of these problems lies in a strategy called encapsulation. Usually, the tool of choice for this is a machine called a spray dryer, which blasts a fine mist of liquid flavor and a protective coating such as starch into a heated chamber to yield fine particles of flavor enclosed in a dry shell of starch. Mary McKee, one of Givaudan's flavor-delivery specialists, shows me a more sophisticated version called a fluid-bed dryer, which suspends the mixture in a strong updraft of air to keep the granules from clumping as they dry. Right now, the machine has lime-green granules bumping up and down in it like bread crumbs in a blender.

McKee, a tall, slender woman whose large eyes seem even larger because of her wraparound safety glasses, opens a port on the machine and dumps a little pile of the granules into my hand. They taste vividly of lime—partly because of the flavor, partly because the color provides a congruent visual cue, and partly because of another trick of the delivery. "When you taste a lime flavor by itself, it's very terpeney with some top notes. We can spray dry that, and it's fine," she says. But in the real world, lime has acid, too, so she spray dried the lime flavor on to crystals of citric acid. Now her flavor granules deliver not just the flavor of the lime, but its citric puckeriness as well. The possibilities here are almost endless. "If we were to spray the same flavor on salt, it would taste very different," she says. Margaritas, anyone? As another example, McKee pulls out a vial of dried oregano leaves coated with jalapeño flavor. Or you could use the same approach to flavor tea leaves. "You can basically coat anything that you can fluidize," she says.

Another technology, which Givaudan has patented, lets them load flavor inside an insoluble capsule without spray drying, and therefore without risking damage to volatile flavorants during heating. The capsules shear easily when you rub or chew them, releasing their flavor intact. Flavors protected like this are perfect in something like the breading on chicken, says McKee, because you can fry the chicken without losing the flavor during cooking. In fact, liquid garlic flavor encapsulated in this way would deliver the same flavor impact as six times as much unencapsulated flavor—a huge cost savings to the producer.

Once a new flavor is finished, the company can move on to the last step of the product development process: testing the final product on consumers. Testing panels actually come in two different sorts, as different as the apples and oranges (among other things) that they evaluate: consumer panels and trained panels. The most straightforward are simple consumer panels drawn from the general public. These untrained panelists are just like you and me—they would struggle for words if asked to describe exactly the flavor of a particular sample. And even if they can find a word, there's little consistency from one panelist to the next. What one calls "fragrant" in the flavor of an apple, for example, another might call "flowery," and a third "sweet." So flavor testers generally don't ask consumer panels to describe flavors. Instead, they stick to simpler questions like "Do you like this?" and "Are these two samples the same or different?"

These are exactly the questions that you want to ask the untrained masses, and Big Food desperately needs to know our answers. Obviously, if you're planning to sell something, you want to know if consumers are buying. Hence, "Do you like this?" and variants such as "Would you buy this?" Even here, though, it's important to make sure you're asking not just the general public, but the right segment of the public. If you're marketing a cheap flavored coffee to be sold at convenience stores, you don't really care what the Starbucks drinkers or the hard-core espresso geeks think of it—you want to ask the folks who actually buy their coffee at 7-Eleven.

Companies often also need to know whether they can cut costs without consumers noticing, so they care a lot about "Same or different?" questions. Unlike the "Do you like this?" question, you can't just ask people outright, because that's an invitation to imagine differences even where none exist—the same over-eager pattern recognition that creates puppy dog shapes in clouds and an image of the Virgin Mary in a grilled-cheese sandwich. Instead, testers give their subjects three samples and ask them to say which one is different—the same triangle test that Joel Mainland gave me when I participated in his "Does this compound have a smell?" study. Sometimes, test organizers use a variant of the triangle test called a tetrad test, in which participants get four samples—two of each—and group them into like pairs. The tetrad test turns out to be much more sensitive than the triangle test, so you need to test fewer participants to be confident of the result.

One winter's day, I got the chance to be part of a consumer panel in the city where I live. I followed directions to a downtown office building, then found my way to the far end of a long, dimly lit hallway not far from the stairwell. Behind a nondescript door that might have fronted a private-eye's office or a low-budget dentist, I found a small, rather austere waiting room containing a handful of other people. Soon the organizers ushered us into the testing room, a row of perhaps a dozen small carrels against an *L*-shaped wall. Behind the wall, I knew, was the kitchen area where staff prepared the samples we were to evaluate.

My carrel had side walls to shield me from seeing what the others were up to, a computer display with a mouse, a cup of water, two saltine crackers, a napkin dispenser, and a bottle of Purell Hand Sanitizer. On the back wall of the carrel was a pass-through with a hinged cover that soon opened to reveal a tray containing a numbered plastic cup (#553) with some pieces of roasted red pepper inside. Aha. I guess we're tasting red pepper.

The computer screen lights up with a question: "How much do you like #553 overall?" It offers a nine-point scale ranging from "Dislike extremely" through "Neither like nor dislike" to "Like extremely." I take a bite. The sample's not bad, so I pick seven, "Like moderately." Then the computer asks, in turn, how much I like the flavor, the appearance, and the texture of #553, and finally, whether I would consume #553 again. Then I push the tray back into the pass-through and close the cover (which opens the pass-through on the opposite side, in the kitchen). I nibble a saltine, take a sip of water, and relax until the next sample appears.

The next one, #310, is sweeter—unpleasantly so—and has a slightly bitter, solventy aftertaste. Is this artificially sweetened, I wonder? The third, #617, seems less roasted. The texture is firmer, but the flavor more insipid. #909 is firm, too, but has that bitter/solventy aftertaste again—my least favorite so far. And #480 is the hands-down winner, with the meatiest texture and the richest flavor. With that, we're done. Sitting back and looking around the room, I see that most of the other panelists—who've done this sort of thing before—are already finished and heading out the door, like studio musicians clearing out as soon as the gig's over.

In the waiting room afterward, the chief scientist explains to me that we've been evaluating a new high-pressure treatment designed to reduce spoilage. The treatment extends the shelf life of the peppers, but some tasters complain that it produces a bitter aftertaste. We're testing whether that aftertaste is noticeable, and how long the peppers can be kept before the flavor starts to deteriorate. (Since we're answering two different questions, we can't do a simple triangle or tetrad test—hence the nine-point scale instead.) The panel as a whole got eight different samples—pressure treated or not, and held for two, four, six, or eight weeks—although to avoid taster fatigue, each individual panelist tasted only five of the eight. "Eight would have been too many samples for one person," she says.

The results are likely to be messy. For one thing, every pepper tastes a little different, so a good treatment on a poor pepper can yield the same score as a poor treatment on a good pepper. And we weren't given any instructions on how to use that nine-point scale, so each taster is likely to score the same pepper a little differently. Someone who roasts their own peppers at home, for example, is probably less likely to "Like extremely" these processed ones, compared with a person whose only experience of red peppers is from a jar or can. Still, given enough tasters—usually around eighty to one hundred people—and big enough product differences, the researchers should be able to find the answers they need. Just before I leave, the scientist breaks the code for me. Samples #310 and #909—the two I thought had an unpleasant aftertaste—were both high-pressure treated, while the other three weren't. #480, my favorite, turned out to be the freshest of all the samples I tasted. If everyone felt as I did, that's bad news for their antispoilage treatment.

These simple consumer panels tell companies a lot of what they need to know about their products—namely, whether people like them. That's why consumer panels have become ubiquitous in product testing, whether for foods or cars or laundry detergents. But unlike cars and laundry detergents, where consumers can generally go into more depth about look and feel, when it comes to flavor, language becomes a problem. What one person calls "very bitter" might be another person's "moderately bitter," or maybe even "sour" or "metallic." That's why the people running my panel never asked us to describe the off taste of some of the peppers, only to say whether we disliked it.

To dig deeper into the flavor details, they would have needed a panel of tasters who agree on what terms like "bitter," "soapy," and "metallic" mean. And that takes training. Companies that want this higher level of sophistication in their flavor analyses typically convene a small group of people—usually just eight or ten—and put them through several hours of training with standard samples to specify exactly what should be described as "soapy" or "metallic" and exactly how bitter "moderately bitter" is. After the panelists have settled into a standard vocabulary, they can start testing the product.

In the case of my processed peppers, the organizers might have trained an expert panel to reliably assign standard descriptors for bitterness, sweetness, roasted flavors, and several possible descriptors for the off aftertaste that I naively called "solventy": soap, turpentine, and nail-polish remover, among others. Then they could present the test peppers to the panel and learn exactly how the various treatments affected the flavor, which might suggest ways to tweak the process to reduce the problem. The catch, of course, is that trained panels are very specific: panelists trained on the descriptors that apply to roasted red peppers won't have the vocabulary for apples or hamburger patties.

Participants in expert panels such as these quickly learn to speak articulately about the intricacies of flavor. The rest of us can take a page from their book and learn to be more articulate about our own flavor experiences. Most of us are pretty good at talking about colors, because we have a common vocabulary to work with. Given almost any color, non-color-blind speakers of English can quickly assign it to one of eleven basic color categories: black, white, brown, gray, red, yellow, green, blue, purple, orange, or pink. From that starting point, we can then make finer distinctions: Is the green a forest green, a kelly green, or a chartreuse? Does it have a touch of blue in it? (Curiously, while English has eleven basic color terms, many other languages have fewer. Some offer just five (black, white, red, yellow, green-blue), three (black, white, red) or even two (light, dark) terms. Imagine trying to describe the color difference between a Granny Smith apple and a Golden Delicious with just "light" and "dark.")

Experts approach flavors in much the same way, by breaking up the flavor world into a handful of basic categories. Givaudan, for example, has developed its own whole language for flavors, which they call Sense It, that lets their customers and flavorists quickly converge on what they're talking about. The details, as usual, are a closely guarded secret.

Over at Givaudan's competitor FONA, on the other hand, Menzie Clarke happily lays out her own set of ten basic categories: fruity; floral; woody; spicy; sulfury (including onions and garlic as well as most meat flavors, eggs, and many off flavors); acid; green (including herbaceous flavors, but also green apples, avocados, and vegetables like beans); brown (nutty flavors, coffee, chocolate and caramel, honey, maple, and bread); terpeney (resinous flavors like pine and citrus peel); and what she calls "lactonic," a category that includes sweet, creamy flavors and the peachy note that was in the strawberry flavor I made in Brian

Mullin's lab. Other flavorists, especially those at other companies, might have slightly different categories. Mary Maier, for example, includes "earthy" and "starchy" in her basic list of savory flavor categories.

Most of the time, though, flavorists and their clients are working within a much narrower range of possibilities—strawberry flavors, say, or chicken. One of the first tasks in any project is to build a glossary of likely descriptors that might apply to the product in question. For strawberries, for example, FONA's basic lexicon includes fruity, floral, buttery, ripe, jammy, seedy, fresh, cooked, green, sweet, candylike, burnt, oniony, and creamy. A list like this gives tasters a ready-made vocabulary for comparing test flavors—and it's always easier to pick the right term from a list than it is to conjure one out of thin air.

One effective way to organize a frequently used set of descriptors is to arrange them in a flavor wheel. The best example of this is the wine aroma wheel developed three decades ago by Ann Noble, a researcher at the University of California, Davis. (If you're not familiar with it, have a look—it's readily available online.) The wheel has three concentric rings, each with a set of descriptions. On the innermost ring, it lists twelve general categories of wine aromas: fruity, vegetative, nutty, caramelized, woody, earthy, chemical, pungent, oxidized, microbiological, floral, and spicy. Suppose you decide you smell something fruity. Then you move out to the next ring on the wheel, which offers six subcategories of fruitiness to choose among: Is it citrus, berry, tropical, tree fruit, dried fruit, or something else? If you pick tree fruit, the outermost ring offers choices that are more specific still: Is it cherry, apricot, peach, or apple? By helping you narrow down the options, the wine wheel quickly lets you arrive at a specific descriptor that fits the flavor of your wine. The approach works so well that there are now flavor wheels for beer, cheese, Scotch whisky, coffee, cigars, chocolate, honey, olive oil—the list goes on and on. (I'm waiting for the day that an ice cream shop posts an ice cream flavor wheel to help people pick which scoop to order. Do you want a berry, spice, tropical fruit, or caramel flavor? If berry, should it be a red berry or a blue berry? Is the red berry strawberry, raspberry, or blackberry?)

These descriptors generally break the flavor world into categories that correspond to flavors found in the natural world—that is, if flavors were paintings, almost all of them would be landscapes or still lifes, more or less faithful representations of subjects in the real world. But are there also abstract flavors that correspond to nothing in the real world? Perfumers, after all, come up with abstract fragrances all the time, but flavorists have barely even ventured into that genre. When I asked flavorists for examples of so-called fantasy flavors, almost everyone mentioned bubble gum, but they struggled to come up with many other examples. There's blue raspberry, perhaps, and certainly Red Bull—a fantasy flavor that, I'm told, was made intentionally unbalanced, to give the impression of vigor, even agitation. In a sense, too, a generic "meat" flavor is something of a fantasy, since all real meat tastes like something—chicken, if nothing else.

(Jeff Peppet once received a phone call from someone who had the notion to make animal crackers with giraffe-flavored giraffes, lion-flavored lions, and so on. Could Givaudan make the flavors? Um, Peppet replied, we don't know what a giraffe tastes like. No problem, said the guy. Neither does anyone else, so just give me a different, novel meat flavor. Givaudan didn't take the project. But still, Daniher gets a little excited by the prospect of developing new meat flavors. "Why not iguana?" he asks, not entirely in jest.)

The notion of mixing cocktails of flavor chemicals to order, as Givaudan does, is sure to make many people uncomfortable. This widespread chemophobia is a big reason why food manufacturers are so leery of being associated with flavor companies, and therefore why flavor houses like Givaudan are so secretive about their clients' identities. As Andy Daniher and Jeff Peppet put it over lunch in Givaudan's cafeteria, companies don't want to be pilloried for "putting chemicals in our food."

From a scientific point of view, of course, this is silly, because all food is nothing but chemicals. The proteins in your steak or tofu are chemicals. The sugars that form the starches in your organic, local, sustainably farmed whole wheat are chemicals. The scary-sounding isoamyl acetate in artificial banana flavor is exactly the same chemical as the isoamyl acetate in a real banana. If you list all the chemical constituents of a banana or an apple—as one Australian chemistry teacher and blogger has done in fake "ingredient labels"—even a simple piece of fruit can sound pretty daunting. (You'll recall from the beginning of the chapter that a real apple contains at least twenty-five hundred chemicals, and a Jolly Rancher candy just twenty-six. If you want fewer chemicals in your food, you should go for the Jolly Rancher every time.)

Even so, what the flavor industry does runs counter to the sense that most of us have that the more "natural" our food is, the better. Food companies want you to feel as though their jar of spaghetti sauce is "just like mamma used to make." Industrially produced flavors don't sit well with that cozy domestic vision, which is why you'll never see "Powered by Givaudan" on the label the way computers advertise that they're "Powered by Intel." "People want to think that came straight from coffee and milk," said Daniher, pointing to my bottled Starbucks White Chocolate Mocha. "But it's a processed product." The PR problem is even worse if the label has to mention "artificial flavor." As a result, food manufacturers—especially the high-end ones—often insist that flavor houses build flavors that can be called natural.

It's worth taking a moment to parse the distinction between "natural" and "artificial" in the flavor world. In the United States, for something to be called, say, a natural lemon flavor, the chemical compounds in the flavor have to be extracted from actual lemons. Naive consumers might think that means they're getting the full richness of flavor found in the real lemon. But in fact, your "natural lemon flavor" might be nothing more than a single chemical, citral. (If you were getting all the flavor depth of the lemon itself, the label would likely read "lemon juice" or "lemon oil," not "natural lemon flavor.") Citral from lemon peel is chemically identical to the citral made artificially in a chemistry lab. If anything, the artificial version is likely to be purer than the natural stuff, which may bear traces of other compounds that tagged along during extraction. But consumers want natural, so natural is what they get, cost permitting.

One step down from "natural lemon flavor" is just plain "natural flavor," a wording that indicates that the flavor compounds come from an actual plant or animal (rather than being made in a chemistry lab), but not a lemon. Natural vanilla flavor, for example, comes from vanilla beans; vanilla flavor made from "natural flavor," on the other hand, usually contains vanillin, the main flavor compound, that's been extracted from wood pulp. (The presence of vanillin in wood is why you'll find vanilla notes in barrel-aged chardonnay or whisky.)

From a scientific point of view, these distinctions are much ado about nothing. Citral is citral, whether it comes from a lemon or a lab. Vanillin is vanillin (though extract of real vanilla beans also contains other flavor compounds that add extra richness not found in synthetic vanillin or the stuff extracted from wood pulp). And a strawberry dessert industrially flavored to simulate ripeness isn't necessarily less healthy than the same dessert made with naturally ripened strawberries that contain the same flavor chemicals. Sure, the strawberries do contain fiber and some other nutrients, but as far as safety is concerned, it's probably all good—at least in the short term.

There might be a deeper problem, though, that goes beyond the safety or palatability of individual chemicals. In the previous chapter, we saw how the body depends on flavor cues to select a nourishing, nutritionally balanced mix of foods. Some critics say that adding extra flavor chemicals to foods tampers with this finely evolved system and thus prevents our bodies from making wise nutritional choices. In essence, the flavor industry is marketing nutritional deceit, they argue, and this deceit contributes to the modern epidemic of obesity and poor nutrition. The journalist Mark Schatzker dubs this "the Dorito effect" in his book of the same name.

When I put this charge to Peppet and Daniher over lunch, they noted that they're only delivering what consumers want. "There's a little chicken-and-egg question," says Peppet. "On the one hand, the food companies are bad because they get people to eat all this salt and fat. But on the other hand, the public wants salt and fat. There's a question of what's driving what."

And besides, they said, there's another side to the story. Added flavors don't have to be bad. "If our customers are willing, flavor can help drive healthier products," says Daniher. People now have the option to choose flavored, unsweetened waters—which use flavorings our brains associate with sweetness—instead of sugar-laden soft drinks. Some yogurt manufacturers have also reduced added sugar by 40 percent by substituting flavorings, instead. "That, to me, is a positive use of a flavor," he says.

In today's world, designing flavors is almost exclusively the task of professional flavorists, plying their trade in secret within commercial flavor houses or other large food companies. But if one visionary Frenchman has his way, within a few decades we may all be concocting flavors in our own kitchens from chemical raw materials.

If you ordered up a mad scientist character from Central Casting, you'd probably end up with someone who looks like Hervé This: about sixty years old, with a tonsure of longish, unkempt gray hair, wearing a white lab coat with the collar sticking up in the back, and an air of earnest, intense—though slightly distracted—enthusiasm. But that crackpot exterior belies This's iconic stature in the food world. He's a household name among avant-garde chefs, a highly respected food scientist, and director of the food division of the French Academy of Sciences. Oh, and he's also the man who coined the term "molecular gastronomy"—the application in the kitchen of precise scientific techniques and ingredients normally found in the laboratory—which has become the hottest field in culinary artistry.

But molecular gastronomy is last year's obsession. This (it's pronounced "teece," by the way) has moved on to a concept far more radical in scope: building foods not from plants and animals, but from what he calls "pure compounds" such as powdered proteins and sugars, and assembling custom-designed flavors from individual molecules, just as Brian Mullin does at Givaudan. He calls the approach "note-by-note cooking," by analogy with a composer assembling avant-garde music note by note from a synthesizer. "For note-by-note cooking, no meat, no vegetables, no fruits, no fish, no eggs," This says in a BBC news report. "Only compounds, and you make the dish."

In part, This thinks the world will be driven by necessity to cook this way. As the world's population rises and fossil fuels and fertilizer become scarcer and more expensive, farmers may find it difficult to grow enough normal food—chicken, cabbage, and rice—to meet the demand. But the things we think of as inedible—This likes to flourish a handful of grass clippings from his lawn—are full of nutritious compounds like proteins and sugars, if only we could get at them. So why not extract the pure compounds and use them as ingredients? You get the extra benefit of a longer shelf life and, perhaps, energy savings from shipping powdered ingredients instead of fresh ones, which are mostly water. (Some skeptics, however, question whether it takes more energy to extract and dry the pure compounds than you'd use in shipping the fresh, wet originals.)

There's a positive side to This's vision, too. Why restrict our culinary flavor palette to the particular combinations of flavors that nature happens to have packaged together? "If you have beef and carrots, you can eat beef and carrots," he told one reporter. "But if you have the 400 compounds in beef and the 400 compounds in carrots, you can make 160,000 combinations. It is like the infinite possibility of making colour from the three primary colours."

Is it realistic to expect that you or I, alone in our kitchens, could actually mix and match pure compounds to make note-by-note dishes in this way? After all, professional flavorists need years of full-time training to understand how to combine flavor molecules into a convincing final product, whether they're mimicking something from the real world like a strawberry or inventing something never tasted before, like Red Bull. The rest of us can't hope to match that level of sophistication. We'll have to start with baby steps: a few bulk compounds for nutrition, rounded out with a simple set of flavor molecules. Can we concoct something tasty from those rudiments, or is mere satisfaction of hunger the best that note-by-note cooking can offer?

I figure there's only one way to know for sure: try it out for myself. A little searching turns up a handful of note-by-note recipes, either from This himself or from an annual note-by-note cooking contest sponsored by his university, AgroParisTech. I'll try one of This's basic recipes, a flavored protein pancake that he calls a "dirac." (One of This's quirks is that he likes to name his dishes after famous scientists, in this case the Englishman Paul Dirac, who predicted the existence of antimatter.)

I'm no test chef, so to give the recipe the best chance of success, I also enlist the help of a real pro: Maynard Kolskog, a research chef and instructor at one of Canada's most highly regarded culinary programs at the Northern Alberta Institute of Technology, just a few miles from my house. Kolskog has a special interest in cuisine that pushes the boundaries, and he's long been an admirer of Hervé This, so he seems eager to sign on to the experiment even though we've never met.

First up, the dirac. This's recipe is simple: three parts what he rather unappetizingly calls "coagulating proteins"—powdered egg white, gluten, pea protein, basically anything that will set up when cooked—mixed with two parts water and some oil and flavored to taste (colored, too, if you like—This favors a bright pistachio green), then fried like a pancake. I meet Kolskog in his research kitchen, where he's assembled the needed ingredients, and we get started.

Our first version, using This's recipe calling for a 3:2 ratio of powdered egg white protein and water, makes a pallid pancake so dense and stiff that after we fry the thing, Kolskog can't even cut it with a metal spatula. "Ooh, that's dreadful," he says. It makes me think of a yoga mat; Kolskog likens it to the weather stripping you'd use to seal a window. For our second attempt, Kolskog suggests more water—a lot more water—and more oil beaten into the batter. He also adds some sugar to the mix. This time, we end up with a light, frothy batter that fries into a much fluffier pancake. "It's better, eh?" says Kolskog when we taste the result. "It's almost edible. That has a little potential." He can imagine using the dirac as a bed on which to rest a slice of smoked salmon, or something equally flavorful.

But by itself, the dirac is a little boring—at least partly because the flavor is so simple. We developed some mildly interesting Maillard flavors as the pancake browned, especially after Kolskog added a little more sugar in the second version. But the main flavor ingredient didn't deliver what I'd hoped. We'd opted for one of This's favorite flavorings, a compound called 1-octen-3-ol, or mushroom alcohol. I love mushrooms, so I was looking forward to the result. Unfortunately, on its own the mushroom alcohol's flavor made me think not so much of mushrooms but of a forest floor on a rainy fall day. Pull aside the top layer of intact leaves to expose the moldering, decaying stuff underneath. That's 1-octen-3-ol. If it smells like mushrooms, they're rotting ones.

The mushroom alcohol would no doubt have worked well as a minor note in a more complex flavor. But now we're back to the same old problem of expertise—to make a worthwhile flavor, I'd have to combine at least several—possibly many—compounds, and I just don't have the training or experience to do that. That's the big advantage of working with real fruits, vegetables, herbs, and meats: a strawberry, or a fillet of salmon, comes ready loaded with a complex mix of flavor compounds—a mix, moreover, that we've already learned to like.

Still, there's no reason a cook like me couldn't gradually learn more sophisticated flavorings. As a start, a little bit of This's pure-compound approach might be worth working into my existing recipes. Now that I've got a bottle of mushroom alcohol on hand, for example, a few drops might add an interesting dimension to the flavors already present in a venison stew. A drop or two of limonene could add a fresh, citrusy lift to a cream sauce or a hollandaise. In fact, tweaking nature, rather than replacing it entirely, is what got This thinking about note-by-note cooking in the first place. He'd noticed that adding a few drops of vanillin—the principal flavoring in vanilla beans, and a key part of the flavor added during barrel aging of booze—made a cheap whisky taste like a more expensive one. (I tried that, too, and I'm not convinced. Maybe I needed a cheaper whisky. A better bet, for my money, would be a drop or two of smoky 4-ethylguaiacol.) An interesting technique, then. But the future of food? Nah. At least, I hope not. I still prefer the real thing, for the most part. And that's where we're headed next—to the farm, to see how our food acquires its flavor.

Chapter 7

THE KILLER TOMATO

Near the southwest fringe of the University of Florida's campus in Gainesville sits a nondescript single-story building of whitewashed brick. A long walk from the massive football stadium and the modern, glass-and-steel high rises of the medical center, the little building looks like it could house a maintenance shop for the campus groundskeepers, or perhaps a storage area for recyclable trash. But for anyone who loves the flavor of a good tomato—and who doesn't?—this could be the most important building on campus.

The supermarket tomato is the poster child for the failure of modern agriculture to produce food with decent flavor. Picked too green, shipped and gassed, the pale pink, styrofoamlike spheres are a faint echo of the sweet, juicy, luscious fruit they could be. Just ask anyone with a sunny backyard tomato patch or access to a good farmer's market. Everybody likes to complain about what's happened to commercial tomatoes. But in that little building in Gainesville, Harry Klee is actually doing something about it. Klee, a horticultural scientist who's spent the past decade and a half trying to uncover the secrets of tomato flavor, knows exactly what's wrong with the supermarket tomato—and he knows how to fix it. Someday in the not too distant future, thanks to Klee, all of us could be enjoying much tastier tomatoes, even from the supermarket, without having to pay a fortune for them.

In his office off the main lab room, Klee—a tall, graying man with a long, thin face, flyaway eyebrows, and a slight cast in his left eye—explains that tomato flavor has been sacrificed to tomato breeders' success in boosting yields, because growers get paid for yield, not for flavor. "Breeders have developed modern varieties that basically yield too much," he says in his pleasant tenor voice. "Think of the leaves as factories that produce sugars, and think of the fruits as consumers. Since 1970 versus today, the modern variety yields 300% more. That's a lot. What the breeders have done, they've made plants that are producing so many fruits that the leaves can't keep up with the fruit." As a result, he says, modern commercial tomatoes are starved of the ingredients that make a tomato tasty: sugars and the volatile odor compounds that deliver a rich tomato flavor. "These modern varieties are literally sucking all the nutrients out of the leaves, and they still can't get enough. So the modern varieties have less volatiles, less sugars, less acid, everything. What's in the modern fruit? Water. So the varieties that the modern consumer is given just don't have the capacity to taste as good as an heirloom grown in your backyard." At first sight, the problem looks intractable. The only way a plant can afford to put more sugar and volatiles—that is, more flavor—into each tomato is to make fewer tomatoes. Flavor and yield, it seems, are on opposite sides of the seesaw: if one goes up, the other must come down. Or must it?

Tomatoes are just one of the many crops that don't taste like they used to, at least in popular belief. But where Klee and a few other researchers have clearly shown that modern commercial varieties of tomatoes lack the flavor of older heirlooms, we know much less about other crops. In fact, there's precious little hard evidence to show that most fruits and vegetables actually did taste better in the past.

If anyone should know, it's probably Alyson Mitchell, a food chemist at the University of California, Davis. UC Davis is smack in the middle of California's Central Valley, where a huge share of America's fruits and vegetables are grown, and it's been a mecca for agricultural research for a century. Yet long-term studies of flavor just aren't available. "There's a lot of speculation—and probably rational speculation—that we are not growing foods with the same flavor," says Mitchell. "It doesn't take a rocket scientist to understand this. When I was a little girl, here in California, we used to go out into the field and pick peaches. And those peaches tasted so delicious. As time goes on, and we buy peaches in the grocery store, the flavor and aroma impact is just not the same. But if you ask my daughter 'What does a peach taste like?' she doesn't have the same historical memory of what a peach tastes like. We don't have that library. The data's just not available to make those kind of comparisons."

Even so, we know that breeders of many crops have focused for decades on traits like disease resistance; yield; appearance; uniform size; and ease of packing, shipping, and processing—all the traits that make the crops easier to grow and deliver to distant markets. Their focus hasn't been on flavor. As one horticulturist told me, kiwi fruit are considered "good quality" if they're the right size and free of blemishes. Flavor doesn't even enter into the equation.

Despite the lack of good, scientific studies that measure crop flavor directly, there might be a backdoor way to document its decline. Fruits and vegetables that are more nutritious are also likely to be more flavorful, because at least some of the molecules that make them more nutritious—the antioxidants in leafy greens, for example—are volatile, or break down into flavor volatiles. Comparisons of the nutrient content of foods over time are easier to find, and they do indeed show that nutrient levels in modern crops, by and large, are as much as 40 percent lower than they used to be. Not every nutrient shows the same decline, and not every nutrient has a direct effect on flavor. However, the overall trend is hard to ignore.

The industrialization of agriculture must also share some of the blame for the tasteless stuff in grocery stores. The peach or cantaloupe in my grocery store in wintry Canadian February has traveled thousands of miles to get there, and to survive the trip, it was almost certainly picked before it was perfectly ripe. This premature harvest cost it the chance to get the full load of sugars and volatiles that a full-term fruit could have acquired. For most fruits, which don't continue to make sugars after harvest, there's no way of making up for the loss. Even in August, when supply chains are shortest, many large-scale producers can't afford to handle fruits carefully enough to let them ripen fully on the tree or vine.

But scientists like Klee are finding ways to put the flavor back into our fruits and vegetables. You'll recall that some aromas—vanilla, for example, or strawberry—can make a sugar solution taste sweeter. If some of the volatiles in tomatoes can pull off the same trick, Klee thought, then maybe growers don't have to sacrifice flavor for yield. He gathered up a wide variety of tomatoes—152 in all, mostly heirloom varieties but including commercial ones as well—and measured the amount of sugars and flavor volatiles that were present in each. The varieties differed enormously, with some volatiles varying by as much as three thousandfold from one variety to another.

Klee chose sixty-six varieties with very different sugar and volatile profiles and, working with Linda Bartoshuk, fed them to a taste-testing panel made up of ordinary people from the Gainesville area. For each tomato, the tasters rated its sweetness, its aroma, its tomato intensity (which Klee defines as "that concentrated 'Wow, that's a tomato!'"), and a few other attributes. They also rated how well they liked that tomato on a scale from negative one hundred to positive one hundred, with the endpoints being the worst and best thing they'd ever experienced. "Effectively, the tomatoes go between 0 and 35," says Klee. "Thirty-five would be a fabulous tomato. Zero is absolutely neutral." (I think I've eaten a few zeros on fast-food burgers or in February salads.)

The panelists generally liked the sweetest tomatoes best. But when Klee looked closer, he saw something more interesting: The level of sweetness that tasters perceived sometimes didn't have much to do with the amount of sugar actually present. Tasters thought a variety called Matina, for example, was about twice as sweet as the Yellow Jelly Bean variety, but the analysis showed that Yellow Jelly Bean actually contained more sugar. Matina tastes so sweet, despite its low sugar content, because it's rich in volatile odor compounds such as geranial that make our brains think "sweet." (Geranial, by the way, is derived from lycopene, the molecule responsible for a tomato's red color. Orange- or yellow-colored tomato varieties make less lycopene and hence less geranial—so they taste about 25 percent less sweet than redder varieties. Something to keep in mind when you're buying tomatoes.)

It's worth taking a moment here to examine why plants have these volatile aroma molecules in the first place. The volatiles that account for the flavors of the plants we eat are what botanists call "secondary metabolites." The term refers to the fact that, for the most part, they're not absolutely essential to the life of the plant, as molecules like chlorophyll, sugars, proteins, or DNA are. Instead, these secondary metabolites serve more subtle functions, often in defense or signaling, or are merely by-products, molecular garbage left behind as the plant performs some other biochemical task.

"Usually, I can best explain what a secondary metabolite is by comparing with humans," says Kirsten Brandt, a plant scientist at Newcastle University in the United Kingdom. "In humans, the main secondary metabolite we have is melanin, a brown pigment. Most people have it in their hair—if you don't, you're blonde—and most of us can make it in the skin. It protects the skin from UV light. Plants make lots of chemicals that they could actually survive without, but they need them to interact with the world around them."

Often, those secondary compounds are there to defend the plant against predators. The bitter taste of broccoli and mustard greens comes from molecules called glucosinolates that are poisonous to many animals, particularly insects, that might otherwise munch on the plants. They're not especially toxic to humans—we dodged that one—but even cattle are more sensitive to the chemicals, which is why canola breeders have made low-glucosinolate varieties for growers who want to feed them to cattle. Similarly, most of the pungent flavors we know from culinary herbs are actually pretty effective deterrents to feeding. (When's the last time you sat down and ate a plateful of rosemary or sage?)

Fruits, on the other hand, want to be eaten. The whole point of a tasty, sugar-filled fruit is to tempt some animal into eating it, carrying the seeds away to be dropped somewhere they won't compete with the parent plant. To help achieve that end, plants endow their fruits with a suite of volatile chemicals that shout out, "Good stuff here! Come and get it!" As Klee notes, many of the flavor compounds in a fruit like the tomato are closely related to essential human nutrients such as particular fatty acids and amino acids that our bodies can't make on their own. That makes these flavor compounds a cheat-proof signal of the fruit's nutritional quality: the plant can't make the flavor compounds without having the nutrients, as well.

The fact that fruits want to be eaten, but only once their seeds are ripe, also explains why "ripeness" only applies to fruits, not to vegetables. Immature fruits contain sour acids and astringent polyphenols—think of an unripe apple, or an immature persimmon—that discourage their consumption; as the seeds mature, the chemical content of the fruits changes from discouraging to encouraging. Vegetables, on the other hand, are always trying to dissuade you from eating them, so ripeness isn't an issue.

But both fruits and vegetables have a common interest in having a recognizable flavor. Remember how we use flavor to learn which foods we want to eat and which to avoid? This is the other side of the exact same coin. Vegetables want us to remember them like Dana Small remembers Malibu and 7UP: that was awful, and I never want to ingest that again. Fruits want us to remember them for their good consequences—except, maybe, for the seeds themselves. The seeds of the coffee plant, for example, pack a potent neurotoxin, caffeine, that most of us are familiar with. In nature, far from espresso machines, this poison teaches an important lesson. "We can learn that we should not eat that plant, because it makes us giddy," says Brandt. "But we need to be able to recognize that plant. It's important to us, and to the plant, that we should recognize the taste." So the coffee plant has evolved distinctive-tasting seeds, and we (that is, mammals) have evolved the taste and odor receptors to recognize those distinctive flavors. After millions of years of this coevolution, says Brandt, "you're not in doubt, when you're eating something, whether it's pea or potato or broccoli. All those past defences are now sitting around being useful for us—and the plants—as labels, to point us in the right direction." Even flavor compounds that today have no toxic effect at all may well have arisen as toxins sometime in the distant past, she notes. And the pressure to evolve new compounds—new flavors—is ongoing. "Anything the plant has been using for a while, their enemies will have evolved countermeasures. So you have an arms race."

Thanks in part to that arms race, tomatoes have at least four hundred volatiles in their fruits. However, only about two dozen are important to the flavor of the fruit, Klee finds—and the important ones aren't necessarily those that are easiest to smell. Until recently, tomato scientists had always winnowed the hundreds of volatiles by comparing their concentration to people's measured detection thresholds. Those compounds whose concentration soared well above threshold, they assumed, must be the most important, while those that fell below the detection threshold could be discarded as unimportant. But when Klee actually tested what made for a tasty tomato, he found that that obvious assumption didn't hold. Some of the most prominent volatile odorants, such as the classic "tomato stem" smell you get when you brush against a growing tomato bush—and which always brings me back to happy memories of the backyard garden—make no difference to whether people like the tomato. On the other hand, some volatiles that turn out to be really important contributors to flavor are present at below-threshold concentrations. Several below-threshold volatiles, it turns out, can work together to alert the brain to their presence—just like Paul Breslin's rose-sweet chewing gum.

Those volatiles are the secret to sweeter-tasting tomatoes, says Klee. It takes a lot of sugar to sweeten a tomato, so growers can't do it without crippling the yield. That's why today you can buy excellent tomatoes in the store only if you're willing to pay a lot more for them. But volatiles don't cost a tomato plant much—they're present in such small quantities anyway that tomato breeders could crank up volatile levels manyfold while barely denting the yield at all. "All of a sudden, you're doubling the perception of sweetness," says Klee. That should make sweeter, richer-tasting tomatoes possible at a price everyone can afford.

Volatiles, incidentally, are the reason why you should never, ever put a tomato in the refrigerator. A tomato is constantly leaking volatiles into the air (which you can easily verify by sniffing a good, ripe one) and replenishing the loss by making new ones. Chilling turns off the enzymes that make the volatiles—and one of the peculiarities of the tomato, a tropical plant, is that the enzymes stay off, even after you take the fruit out of the fridge. Volatile content goes down as molecules leak out into the air and aren't replaced, so a tomato that's spent time in the fridge tastes less sweet and has less tomato flavor. (And, by the way, most of the volatiles leak out the stem scar at the top of the tomato—so, all else being equal, a tomato sold "on the vine" with a bit of stem attached ought to be a little more flavorful than one without.)

Klee has already taken the first steps down the road to the flavorful tomato of the future. In 2014, his team released its first two new varieties, Garden Gem and Garden Treasure, that were created by crossing high-volatile heirlooms with modern, high-yielding varieties. The hybrids yield nearly as much as the commercial varieties but keep almost all the flavor of the heirlooms, he says. As I talked tomatoes with Klee, five golf ball–sized Garden Gem tomatoes sat between us on his desk. After a couple of hours, he offered to cut them up so I could taste what we were talking about. It wasn't an ideal test—these were April tomatoes, after all, ripened when days were short and temperatures relatively low, so there was little chance they'd reach the glorious heights of a midsummer heirloom. They didn't—but they certainly tasted sweeter and tomatoier than anything I was likely to get in the grocery store at the time. Grown under better conditions, Klee's two new varieties have definitely gotten people excited. As I write this, the varieties are not yet commercially available, but Klee has sent seeds to more than thirty-two hundred people who donated money to the tomato research program, and the feedback is enthusiastic. "We've had several people write to tell us they're the best tomatoes they've had in their lives, which makes us feel good," he says. "People really want these things. It just shows how big the pent-up demand is for good tomatoes."

Just an hour down the road from Klee's lab, in the sandy soils of central Florida, a plant breeder named Vance Whitaker is trying to solve the flavor problem for another fruit that's often disappointing in the grocery store: strawberries. The big problem with strawberries is that they're what's called a "nonclimacteric fruit," meaning that they don't ripen any further after harvest. You can't treat a strawberry like a banana or an apple or a pear—or a tomato, for that matter—and prod it into ripeness in the warehouse with ethylene gas. All you can do is let it ripen on the bush as long as you dare: once you pick the berry, it's all downhill. It'll never get any better. And because strawberries are so fragile, growers can't risk letting them ripen fully before picking, because they'd never survive the rigors of shipping and handling. The result is that a grocery-store strawberry will rarely be as ripe as one you'd pick yourself—and the clearest sign of that is the white "shoulders" you'll see at the stem end of most grocery-store berries.

What to do? Scientists could try to find a way to help the strawberries last longer, so that growers can afford to pick them when they're riper. Or they could look for a way to boost the flavor directly. Whitaker chose the second route, and started digging deep into the chemistry of strawberry flavor, using the same techniques that Klee used for tomatoes. (In fact, the two research teams share several scientists in common, including Klee himself and Linda Bartoshuk.) Strawberries are a lot like tomatoes, he found: People like sweeter ones better, and they also prefer a more intense strawberry flavor, which depends on the volatile chemicals. And, just like tomatoes, if the plants make too many berries, they can't afford to stock them with enough sugar. That puts breeders like Whitaker in a bind. "We could increase yield by a pretty sizeable percentage in just a couple of generations," he explains, "but we would drastically reduce sugar content."

One solution might be to breed a more vigorous strawberry

plant that photosynthesizes more energy, so that it can afford a bigger sugar budget. A more likely option, though, might be to take a page from Klee's playbook and tinker with the volatiles. Sure enough, when Whitaker and his colleagues looked at the volatiles in strawberries, they found several that made the berries taste sweeter, independent of the amount of sugar that was actually present. (Curiously, even though many of the same volatiles turn up in strawberries as in tomatoes, different ones affect sweetness in the two fruits. It's all in the context.)

Strawberry flavor intensity, too, depends strongly on the mix of volatiles that are present in the berry. Whitaker is looking closely at a molecule called gamma-decalactone—the very same peachy-smelling molecule that bridged top notes and bottom notes in the artificial strawberry flavor Brian Mullin designed for me at Givaudan. Some strawberry varieties have it, some don't. Whitaker's team sorted through the genotypes of the haves and the have-nots—a harder task than it sounds, because strawberries have not two but eight copies of each gene—to find a single gene variant that accounted for the difference. With a clear target identified, breeders will have an easier time ensuring that any new varieties have the good gene for this important flavor compound. They can use the same technique—and all of Whitaker's genotyping work—to find other flavor genes more quickly.

There are a few other, nongenetic secrets to growing tasty strawberries, says Whitaker. Cool temperatures, especially at night, help the plant store more sugar in the berries. As a result, Florida strawberries always taste best early in their growing season, in December and early January; quality declines as the weather heats up into February and March. (It seems counterintuitive to put strawberries at the top of your shopping list during the darkest days of winter, but that's exactly what you should do, at least if your grocer gets berries from Florida. Berries that come from California or Mexico have different seasons for peak flavor.) And just a little bit of water stress, or just a little bit of fertilizer limitation, also tends to improve flavor by slowing growth and giving the plants plenty of time to stock the berries with sugar and volatiles. In contrast, good soils don't seem to make any difference. Whitaker's soils in Florida are nothing but coarse sand, and many growers in Asia and the Netherlands produce delicious berries hydroponically, with no soil at all.

Tomatoes and strawberries are unusual in having crop scientists pay much attention to flavor. Most other fruits, and almost all vegetables, exist in a vast, undifferentiated sea of produce where one head of broccoli is interchangeable with the next. "The guy who's buying the stuff for the supermarket, he wants it to taste the same as last time," says Brandt. "What most of the supply chain wants is predictability and low price. There are no consumers wanting special broccolis. They're just not there." In such a milieu, it's not surprising that the science of flavor on the farm is almost nonexistent.

We don't know much, for example, about how a farm's soil affects the flavor of the crops. (Not much at all, in the case of hydroponic strawberries!) Here's Alyson Mitchell on one of the crops she studies, spinach: "I don't think there has been a single sensory study done on spinach looking at the effect of growing environment on flavor. I would be shocked to find it." And most other crops are in pretty much the same boat. Once again, we can get a little more information by asking about nutritional quality instead of flavor, since that attracts a little more research money, but even there, big lessons aren't easily found.

There is one crop, however, where flavor matters above anything else, even yield: wine grapes. The whole point of growing wine grapes, of course, is to make a wine with a distinctive, appealing flavor. If anyone knows how soils and farming methods affect the flavor of a crop, it's going to be viticulturists. For an example that's been worked out in great detail—an example, moreover, that you can taste for yourself tonight—let's journey halfway around the world from Harry Klee's tomato lab to New Zealand.

Mike Trought loves to talk wine—and there's probably no one in the world who knows more about the acclaimed white wines made from the sauvignon blanc grape, particularly in the Marlborough region of New Zealand's South Island. A cheerful, balding fellow who's spent more than three decades in Marlborough as a researcher, university lecturer, winery consultant, and viticulturist, Trought recalls the time, in the midnineties, when he made a pilgrimage to the world-renowned oenology department of the University of California, Davis. At the time, New Zealand wines were just beginning to emerge onto the world stage, and Trought poured two bottles of Marlborough sauvignon blanc for a group of Davis researchers. Everyone dismissed them, calling them too acidic, too herbaceous, unsubtle—in short, unripe and inferior. Two decades later, the joke's on the folks from Davis. "New Zealand sauvignon blanc has now become a benchmark for sauvignon blanc around the world," says Trought. "We can't produce enough." One of the wines Trought poured at Davis that year, Cloudy Bay, quickly became so popular that wine shops couldn't keep it in stock, especially in Britain. (Trought wonders, a bit impishly, whether wine experts might actually hinder progress in the wine world.)

If you're at all fond of wine, you have probably encountered the distinctive flavor of New Zealand sauvignon blanc. Sip one of those wines, especially from Marlborough, and you'll be struck by the pronounced aromas of passion fruit, green pepper, and what's often described as boxwood, or "cat's pee on a gooseberry bush"—the latter being an actual, if unconventionally named, commercial wine. The intense, distinctive flavors make New Zealand sauvignon blanc an ideal test case for understanding where a wine's flavor comes from and how growers and winemakers can influence the outcome. As an added bonus, New Zealand's wine industry is relatively new, so tradition doesn't get in the way of science.

What gives Marlborough wines their distinctive flavor? It's certainly not the grape variety alone. Virtually all of New Zealand's sauvignon blanc vines are descended from a single clone originating in the vineyard of France's fabled Chateau d'Yquem, where they yield a vastly different wine. Instead, a lot depends on the vineyard soils. Not, perhaps, in the way you'd think. The notion that you can somehow "taste the soil" in a wine is completely false. Grape vines take up only water and simple nutrients like nitrogen, potassium, and calcium from the soil. They make all their more complex biomolecules—including the flavor volatiles—in-house. To put it more bluntly, none of the volatile molecules that determine a wine's flavor come directly from the soil. (This calls into question one of wine writers' favorite buzzwords these days: "minerality." You won't find the term much in wine writing before the 1980s, but it's become a term of high praise today. Whatever the term means—and the experts don't exactly agree—it's not the flavor of the vineyard. One study suggests that "minerality" is a description that emerges only when a wine lacks any other distinctive flavor.)

Instead, a vineyard's soil affects flavor indirectly, by altering how the vine grows and, especially, how quickly the grapes ripen. In Marlborough's Wairau Valley, the vineyards sit on an old river flood plain, where the soil is a jumble of sand, gravel, and cobble deposited by the river channel as it meandered across the plain. Soil quality can change rapidly as you walk through the vineyard, so that grape vines only a few yards apart experience very different soils. Where the soil is shallower, vines tend to be less vigorous, and their fruit ripens earlier. (Trought isn't sure why the smaller vines on stony soils ripen earlier, but he suspects it might be because the vines put more of their energy into the grapes when growing conditions are poorer.) When harvesters go through the vineyard, the patchwork of soils means that some bunches of grapes will be harvested at a riper stage than others. The volatiles responsible for the green pepper flavor, known as methoxypyrazines, form early in grape development, so they are more prominent in less ripe grapes. Meanwhile, the thiols that account for the passion fruit notes dominate in riper grapes. This mix of ripeness, and the different flavors it delivers, helps give the Marlborough wines their complexity. "To some extent, that's the characteristic of Marlborough sauvignon blanc," says Trought.

There's more to the story, though, as Trought and his colleagues discovered. "When we started our sauvignon blanc program, we did it because we thought it was going to be easy," he says ruefully. "As we got into it more and more, we realized it was much more complex. It's not just what's in the vineyard that matters." If you pluck a grape off the vine and chew it, you won't notice much passion fruit flavor, because the thiol molecules haven't formed yet—only their odorless precursors are present. The thiols themselves form during fermentation, as the yeast attack the precursors and split off thiol molecules. Rough handling of the grapes causes them to accumulate more of the precursors, so machine-harvested grapes yield wines with about ten times as much thiol as handpicked ones. This, incidentally, may be part of the reason that New Zealand sauvignon blanc, which is generally mechanically harvested, tends to have a much more pronounced passion fruit flavor than French sauvignon blanc, which is usually hand harvested. Even trucking the grapes from vineyard to winery leads to more thiols in the finished wine.

The biggest effect on the final flavor of a wine comes from fermentation, as wine yeasts and other microbes attack the sugars, proteins, and other molecules in the grape juice and convert them to alcohol and flavor volatiles. Each strain of yeast approaches this task with its own unique tool kit of genes and enzymes, and as a result different yeasts can yield very different wines from the same juice. Winemakers are very aware of this, and put a great deal of thought into their choice of yeast. Here, too, regional differences matter, because every winegrowing region—and, quite possibly, every vineyard—harbors its own unique microbial ecosystem. Winemakers rarely sterilize their grapes before fermentation, so these microbes end up in the fermentation tank. In fact, many winemakers rely exclusively on natural microbes for fermentation. So it makes sense that part of the regional character of a wine—its "terroir," to use the term beloved of wine critics—might be the result of different microbial actors taking the stage during fermentation.

Plausible, but until recently, untested. A few years ago, geneticist Sarah Knight and her colleagues at the University of Auckland, New Zealand, set out to see whether it really was true. To rule out any differences in the grapes themselves, Knight started with a single batch of Marlborough sauvignon blanc grapes and sterilized them to kill any resident microbes. Then she divided the juice into a series of tiny fermentation tanks, and sowed each one with a different wine-yeast variant gathered from one of New Zealand's six main winegrowing regions. Same juice, same fermentation conditions—the only difference was the yeast itself. In the end, the yeast variants from each region produced a wine with a detectably different aroma profile. Theory confirmed! Not only that, but Knight's study probably underestimated the effect of microbial differences, because she used only wine yeasts, not the whole microbial flora.

For other crops, too, any effect of the soil on flavor is likely to be indirect. The soil a plant grows in determines how much water and nutrients it has access to, and therefore its energy and materials budget for the sugars and volatiles that determine flavor. You'd think that more would always be better, but it's more complicated than that.

To explain, I turned to Carol Wagstaff, a crop scientist at the University of Reading in England—just a few minutes' drive, actually, from chef Heston Blumenthal's Fat Duck restaurant in Bray. Reading's research group is one of the few to actually study how growing conditions, shipping, and storage affect the nutritional value and flavor of crop plants. Wagstaff has unruly, long brown hair and a large, strong face that lights up when she talks about her work. If conditions are too easy, she says, plants have little need for secondary compounds and put all their energy into growing as fast as they can. Only when they begin to feel a budget crunch do they invest in defending what they've already got. "A bit of controlled stress doesn't go amiss. When a plant is stressed, you'll get more secondary compounds, and that means more flavor and more nutrients," she says. That's likely why Whitaker's strawberries also profit from a little water stress. Exactly what that stress response means in flavor terms is likely to depend on what Wagstaff calls the "metabolic bureaucracy" of the plant—that is, its particular genetic endowment of enzymes and the particular balance of secondary chemicals they favor. Much of Wagstaff's research centers on arugula, which the British call rocket, and that's exactly what she sees there. "You can see quite clearly that some genotypes of rocket will preferentially shunt production in one direction, and other genotypes head down another route when they're stressed," she says.

Soil microbes, too, could play some role in determining the flavor of the crops they grow with. For example, baby corn—a popular vegetable in some Asian cuisines—contains a volatile flavor molecule called geosmin, the same molecule that gives red beets their earthy flavor. Researchers think the young corn plants don't make the geosmin themselves, because corn grown in an English greenhouse lacks the compound. Instead, they think microscopic fungi living with the roots of the corn plants make the geosmin. The plants take up the fungi through their roots, and the geosmin comes along for the ride. It's possible that soil microbes affect flavor in other ways, too, but so far there's little actual evidence.

So far, we've been talking as though more flavor was always better—but for many vegetables, particularly members of the sharp-tasting mustard family, like arugula and brussels sprouts, that's not necessarily true. Many people—especially those who carry the bitter-sensitive version of the *T2R38* taste receptor—find the bitterness of their secondary compounds off-putting and would prefer that their brussels sprouts have less flavor, not more. "Horticulture is essentially messy," says Wagstaff. "You've got the variable genotypes of your plants, you've got the environment you're growing them in, and you've got the varying genotypes of the consumer."

Once a fruit or vegetable has been picked, its flavor continues to change during storage and en route to your grocery store. Partly, that's because volatile flavor molecules leak out into the air, as we've seen happens with tomatoes. At the same time, though, enzyme activity in the tissues can produce new flavor molecules or alter old ones. Occasionally, this can mean that a fruit or vegetable actually improves with storage. Arugula, for example, continues to produce glucosinolate molecules during cold storage after harvest. When you chew a leaf, they turn into flavorful isothiocyanates. That's good news for your salad: The arugula you buy in the grocery store—if it's relatively fresh—may actually be more flavorful than if you'd picked it in your own garden this afternoon. After a few more days in the refrigerator case, though, that advantage goes away, as the "fresh" set of flavor compounds gives way to nastier products that result from fat breakdown. This happens at different rates depending on the variety of arugula—some store better than others, Wagstaff has found.

Some other vegetables last a long time with little or no change in quality. An onion or a potato, for example, is meant to just sit there like an inert lump—that's its job, as a storage organ for next year's growth. So it makes sense that we don't notice much of a decline in flavor. Others, such as corn and carrots, are sweetest just after picking, because enzymes convert their sugar into starch, and no new sugars arrive after picking. They'll last, but their flavor will be disappointing. But a head of broccoli or an asparagus spear hasn't evolved to be long lasting. Far from it—both are rapidly growing shoots, and as soon as you pick them their flavor starts to degrade. One Spanish study, for example, found that more than 70 percent of the glucosinolates in a freshly cut head of broccoli have vanished after a week in cold storage, and another 10 percent disappear after another three days on a grocer's display. That's a lot of lost flavor.

Many people think that another way to ensure tastier fruits and vegetables is to buy organic when possible. It makes sense, in theory: If a little stress is good for flavor, then you'd expect that organic crops ought to benefit, flavor-wise, from the extra insect damage and weed competition they experience. Hundreds of scientific studies have compared the flavor—or, more often, the nutritional content—of organic and conventional crops. The results, unfortunately, are a mess. Some studies show that organic crops are indeed better, while others find no difference. Even the so-called meta-analyses—in which researchers scour the library for every comparison they can find, then add up the results to get a majority opinion—haven't reached consensus on whether organic is better.

A big part of the problem is that the answer you get depends on how you ask the question. You could go to the grocery store, buy a head of conventional and a head of organic broccoli, and measure—

or taste—the difference in secondary compounds. But if the conventional broccoli was harvested two weeks ago in Mexico, and the organic was picked yesterday just down the road, that difference in freshness might have much more impact on the flavor than any organic versus conventional effect. It's hard to generalize, though. The Mexican broccoli could have gone straight into a refrigerated warehouse and stayed under refrigeration right until the time you put it in your shopping cart, while the local one could have spent a hot summer's afternoon in the back of a pickup truck, followed by hours in the sun at the farmer's market. In that case, local might not mean fresher.

Ideally, you'd like to compare the flavor of identical crops grown side by side with organic or conventional methods, because that cuts out a lot of the potential sources of confusion. Researchers at Kansas State University did exactly that a few years ago, planting onions, tomatoes, cucumbers, and several leafy greens in greenhouses in identical soils. When the crops were harvested, about one hundred volunteers tasted organic and conventional samples of the same vegetable—without knowing which was which—and rated how much they liked them and how intense the flavors were. The results? It didn't matter one bit whether the vegetables grew organically or conventionally. The taste testers liked them all equally (or, in the case of mustard greens and arugula, disliked them equally—evidently Manhattan, Kansas, is not the place to get rich from an arugula greenhouse). The only difference was that people thought the conventional tomatoes had a little more flavor, probably because they were also a little riper.

That's not to say that you won't find organic produce more flavorful. As we have seen, our expectations play a big role in flavor perception—for example, wine tastes better when we think it's expensive. That bias probably comes into play here as well: If you think organic produce will taste better, then it probably will, to you. Consider what happened when Swedish researchers gave unwitting university students two identical cups of coffee, telling them that one was "eco-friendly" and the other conventionally grown. Sure enough, most volunteers thought the eco-friendly coffee tasted better—and the effect was strongest for people with the strongest environmental consciousness.

Even if organic farming or other differences in growing conditions do make a difference to crops' flavor, it's likely to be less important than flavor differences among varieties. If so, then breeders, not farmers, may be the critical link in producing tastier fruits and vegetables. In upstate New York, for example, Cornell University plant breeder Michael Mazourek has been working on breeding a more flavorful squash. Squash and other vegetables are the poor cousins of the agricultural world, Mazourek says. You can go into any grocery store and find perhaps a dozen different apple varieties, each offering its own recognizable flavor profile. And we know them all by name: a Granny Smith will be tart and firm, a Spartan sweet and softer, Golden Delicious rich in estery fruit flavors. No doubt you have your favorites. But can you name your favorite variety of broccoli, or your favorite butternut squash? I'll bet not.

"Vegetables are still part of a commodity system, where sameness is one of the overarching goals," says Mazourek. "There's not value in people being able to tell that the bell pepper in the grocery store is different from the one last month. It's an anti-value." With all the commercial pressure working in the direction of sameness, there's little incentive for anyone to develop a tastier version.

Mazourek is trying to change that. He starts by seeking out heirloom squashes noted for their good or unusual flavor and crossing them with commercial varieties, then planting the progeny out in his field. When the fruits ripen, he gathers them and selects the most promising ones for taste testing. "I can't possibly eat some of every squash and stay sane, so we have some proxies that narrow the pool that we do the taste tests on," he explains. First, he picks fruits with the highest level of dissolved solids—that is, sugar and other molecules that might contribute to flavor. Then from those he selects the ones with the deepest yellow flesh. Those have the highest levels of carotenoid pigments, key precursors of many flavor compounds. Unlike Harry Klee's work with tomatoes, Mazourek doesn't yet know which of the many flavor molecules are most important in a good-tasting squash, so he can't measure them directly with a gas chromatograph. Mazourek has to do his flavor analysis the old-fashioned way: he roasts several varieties of squash and sees which ones he likes best. (By the way, here's a squash specialist's advice for the most delicious way to roast a butternut squash: halve it and scoop out the seeds, cover and roast in a four-hundred-degree oven for forty-five minutes. Then uncover the squash, baste it with butter or oil, and continue roasting until tender. "It's not what Betty Crocker says," Mazourek admits. "Roasting them for longer and hotter really is a way to bring out a lot of the savory flavors layered on top of the sweetness.")

After many generations of breeding, Mazourek has ended up with what he thinks is the best butternut squash on the planet. Sometimes called the Barber squash—after Dan Barber, a New

York City chef who encouraged his efforts and now serves the squash in his restaurant—Mazourek's squash has more dissolved solids and a higher carotenoid content than any other squash. "Everything is amped up," says Mazourek. Better yet, the fruits have a built-in ripeness indicator, changing color dramatically from deep green to a rich caramel brown when they're perfectly ripe, so that pickers can be sure they spend long enough on the vine. "There's about a fourfold boost to the carotenoid content in the Barber squash," Mazourek says proudly. "Half of that is in the squash itself. The other half is that it's ripe."

For most consumers, though, the biggest immediate payoff is likely to come from Klee and his tomatoes. Since my visit, Klee has been digging deeper into the genetics of tomato flavor. In collaboration with a research group in China, he has now fully sequenced the genomes of more than four hundred tomato varieties, and fully mapped out their chemical content. In the same way that human geneticists search the genome for gene variants, or alleles, that contribute to diseases, Klee has scoured these tomato genomes for alleles that are important for production of sugars and volatiles. And he can compare modern commercial varieties to heirlooms to see exactly where breeders went wrong.

Back in the 1920s, breeders latched on to a chance mutation that eliminated the dark green "shoulders" on unripe tomatoes. The new, uniformly colored fruit helped growers decide the best time to harvest, and consumers preferred the solid red color in the grocery store. ("People do buy with their eyes," says one tomato grower.) The mutant looked like a big winner—and, in fact, almost all commercial tomatoes grown today carry this mutation. But there was a downside: To allow the green shoulders to redden fully, the mutation interfered with the production of chlorophyll in the fruit. Less chlorophyll means less photosynthesis. The new tomatoes lost out on a sugar boost that green-shouldered tomatoes enjoy—and, as a result, uniformly ripening tomatoes have about 20 percent less sugar.

For volatiles, the losses were even worse. Over the course of decades of breeding for high yield, the high-producing alleles for flavor volatiles have simply fallen by the wayside, because breeders didn't know they were important and didn't test for flavor. "For the volatiles, I'd say at least half of them are the wrong alleles," says Klee. Fortunately, the good alleles are still there in the heirloom varieties—and now that Klee knows which genes are important, it should be straightforward to breed the good alleles back into higher-yielding varieties. "The road map is very clear. We know exactly what we need to do," says Klee. "It just takes time."

It shouldn't be long before everyone should have access to tastier tomatoes, and perhaps other crops as well. Even before that, though, cooks will still strive to draw as much flavor as they can from the raw materials in their kitchens.

Chapter 8

THE CAULIFLOWER BLOODY MARY AND OTHER CHEFLY INSPIRATIONS

Hyde Park, New York, used to be famous as the hometown of U.S. president Franklin D. Roosevelt. To most food lovers today, it's better known as the site of the Culinary Institute of America. The CIA, as it's known, is America's preeminent culinary school, the nursery that incubated countless top chefs.

For all its august stature now, the CIA had a fairly modest beginning. As the Second World War drew to a close, America faced a flood of newly demobilized young soldiers who needed jobs—and, in many cases, needed job skills, having spent the entirety of their brief adult lives in the military. The wife of Yale University's former president figured some of those returning soldiers could find work as cooks, so she founded the New Haven Restaurant Institute to teach them how. The idea took off, and the little cooking school soon acquired grand ambitions and a grand name to match. By 1970, the CIA had outgrown its home near the Yale campus in Connecticut, and it moved into its present location on the Hudson River, an hour upstream from Manhattan. The big brick main building was once a Jesuit novitiate, a retreat for trainee priests—a nice parallel for today's novices as they prepare for a lifetime's dedicated service at the stove, rather than the altar.

Of all the CIA's eminent instructors, the two best positioned to bridge the gap between the science of flavor and its application in the kitchen are Chef Jonathan Zearfoss and Dr. Chris Loss. Together, Zearfoss and Loss teach flavor science to the aspiring chefs. They've spent a lot of time pondering the science behind what tastes good, and both of them are comfortable in the lab as well as in the kitchen. I met the two over lunch at the CIA's Italian restaurant, the Ristorante Caterina de'Medici. As we sip spritzers of grapefruit juice and mint, Zearfoss explains that much of what a chef does in designing a dish is to balance contrast and similarity of the ingredients. Foods go together, he says, either because the flavors of one echo those in another so that they blend well, or because their different flavors make one another stand out. Every chef navigates his or her own path between those two beacons. It's trendy to serve Hendrick's Gin with cucumber to underscore the cucumbery notes in the gin itself, for example. But Zearfoss—who says he's more of a contrast guy—always asks for lime instead, because its angular acidity contrasts against the roundness of the gin's cucumber notes.

Zearfoss and Loss are their own study in contrast and similarity. Zearfoss is a large, imposing man with a shaved, bullet-shaped head, small eyes, and regulation chef's whites, and he speaks with the authority of someone who's used to having his way in the kitchen. Loss is small, dark, and high-strung, with curly black hair and rapid speech. He's in a suit, but tieless. It's hard to give firm rules for managing contrast and similarity, Loss says. Almost everyone likes contrasting textures—a bit of crunch here, some creaminess there. Good chocolate brings its own internal textural contrast, as you snap off a bite only to have it melt in your mouth; ice cream gives a comparable textural treat. Often if chefs are using an unusual ingredient or presentation, they'll pair this novelty—a contrast, in a way, with our expectations—with other, more familiar ingredients that help settle diners' innate neophobia. But mostly, chefs just have to trust their instincts. "Sometimes it's hard to identify what works," he says. "It's easier to pick out the flaws."

One of their favorite lab exercises is to let students use the principles of similarity and contrast to pair wine with foods. The ideal wine for the job is sauvignon blanc. The similarity angle works because of sensory suppression and release, says Zearfoss. You take a sip of the wine and enjoy the balance of flavors. Then take a bite of green pepper. The pepper contains grassy-tasting methoxypyrazine, which makes your nose less responsive to the similar methoxypyrazine notes in the wine. As a result, the next time you sip the wine, you'll probably notice one of its other flavors instead. The wine tastes different at each sip—a complexity that adds to its interest. Certain foods—pears, passion fruit, grapefruit, and others—will inhibit other parts of the wine's aroma, and give different sensory experiences.

Still, a note of caution is in order. If anyone has a stake in the science of pairing food and wine, you'd think it would be Terry Acree. Acree is a flavor chemist at Cornell University with a wide-ranging intelligence that he loves to poke into the dusty cracks among scientific disciplines. Over the past few decades, Acree has put a huge amount of effort into cataloging the flavor molecules that we encounter in our food, and he's published extensively on the flavor chemistry of wine. Here's what he told me about the principles that determine whether a particular food and wine go together:

> What does it mean to "go together"? My mother was an interior decorator, and when I was about five, I walked in and said to my mother, "My favorite color is red." And she said, "No it isn't, kid. That's the stupidest thing I ever heard of. Nobody has a favorite color. Color has a place, and you have to find out where it belongs and where it doesn't belong. It can only be your favorite if it's in the right context." So the first thing I've got to say about wine and food pairing is that it's completely contextual, and almost entirely individual. It makes no sense to write a book on wine and food pairing, except to say there is such a thing as wine and food pairing, and go figure it out for yourself, because it's your own pairing that counts.

As I'm talking wine with Zearfoss and Loss, the servers—CIA students getting some front-of-house practice, and clearly a bit nervous with the chef and professor at the table—bring our food. Zearfoss has ordered vitello tonnato, cold poached veal covered with a tuna sauce, while Loss is eating steak and fries. Loss pushes his fries into the center of the table for everyone to share. (Too many rich, delicious meals are clearly an occupational hazard to be resisted—both men ordered with restraint and ate abstemiously.) Zearfoss eats a fry and then gestures toward his vitello tonnato. "They should have served this with french fries. It's the perfect combination." He points at the uniform, soft beige of his dish. "There's no brown, there's no crunchy, there's nothing here with a Maillard." In short, not enough contrast for his taste.

The french fries come with a little crock of mayonnaise for dipping, which carries another lesson in balance. Without the mayo, the fries come across as too salty; with the mayo, they're just right. "You don't get the same salt impact with the mayo, because the fat coats the tongue," says Zearfoss. "That's the challenge of a chef. You've got the salt, the potato, the fat. Ultimately, what you're trying to get in the customer's mouth is a combination."

Every creative chef approaches the challenge of balancing the flavors in a dish in his or her own way. Many think in terms of base notes, middle notes, and top notes, just like industrial flavorists do. A French onion soup, for example, might have base notes of the oniony flavors, middle notes of caramelized sugars from long cooking of the onions, and a top note of sherry vinegar to make the whole dish sing. Other chefs free-associate from one flavor to the next, building the finished dish in their imagination until it's just right. There aren't many common threads here, as a perusal of any collection of famous-chef cookbooks demonstrates clearly.

Where we can find commonalities, though, is in the chemistry of cooking. In a sense, a cook's job is to collect and curate the right set of flavor molecules.

The first way cooks can intensify flavor is by extracting and concentrating aromatic molecules to deliver a more intense hit of flavor. Extraction is all about solubility: most flavor volatiles, such as the terpenes responsible for rosemary's piney quality, are more soluble in oil than in water. As a result, if you toss a handful of rosemary into a stew, relatively few of the terpenes will end up in the liquid; instead, they vaporize into the air, making your kitchen smell delicious but doing nothing for the stew itself. Better to fry the rosemary in butter or oil first, along with onions and garlic, so that the terpenes extract into the oil and stay in the dish. Or buzz the herb in a blender with a little oil, then strain out the leafy bits, for an intensely flavored rosemary oil to drizzle over the stew at the table.

On the other hand, sometimes you want to minimize extraction to keep as much flavor as possible in the food itself—especially if you're going to discard the cooking liquid. Some of the key flavor molecules in asparagus, for example, are water soluble, so if you boil asparagus, they extract into the water and end up down the sink. Sauteeing the asparagus in butter or oil minimizes this loss and keeps more of the flavor in the vegetable. For the same reason, broccoli and beans—whose key odorants are oil soluble—retain their flavor better if steamed or boiled.

In high-end professional kitchens, chefs can concentrate flavor molecules extracted from herbs, spices, or almost anything else—including soil, seawater, and vegetation—using sophisticated (and expensive) distillation apparatus. Most of us lack the machinery to do that, but anyone can concentrate flavors via simple evaporation—for example, when we cook down a wine sauce to a syrupy consistency. Even the trainee chefs at the CIA quickly learn to have a cauldron of stock slowly reducing on the back of the stove. The process inevitably loses some of the flavor to the air, as a quick sniff will verify, but the reduced stock still packs a more intense flavor.

The second way cooks develop flavor in the kitchen is through cooking itself—the application of heat. The transformations that heat performs on flavor are largely a matter of breaking down big molecules such as fats and proteins into smaller, more volatile ones. This is most obvious with meat, so we'll take that as an example. In the raw state, most meats have relatively little flavor. Anyone who's eaten steak tartare or sushi knows how subtle, almost minimalist, the flavors are. In fact, you probably wouldn't be able to tell much difference between cubes of raw beef, lamb, and pork. All have a mild flavor that's often described as "blood-like," and a slight tang of iron. The vegetable world is diverse: sometimes we eat flower buds, sometimes leaves, sometimes roots, and sometimes fruits, and they carry a wide range of volatile molecules as attractants and as chemical defenses against marauding herbivores. By contrast, most of what we call "meat" is the muscle tissue of mammals or birds, and every one of those muscles is doing roughly the same thing with roughly the same set of biochemical tools. That's why beef and lamb taste more like each other than beets and broccoli do.

The difference between one meat and the next is mostly a matter of the fat molecules they contain, with beef containing larger, less branched fat molecules and lamb, pork, and chicken increasingly more of the shorter, branched molecules. These fats—to be more precise, we should call them fatty acids—differ mildly in flavor on their own, but they break down into much different flavor molecules during aging and cooking. The fat that makes the most difference here, incidentally, isn't the visible fat tissue on and between muscle fibers, the stuff you trim off if you're meticulous about counting calories. Instead, most of the distinctive flavor of lamb, beef, and pork comes from the fat molecules known as phospholipids, which make up the membranes that enclose each cell. Researchers in England demonstrated this more than three decades ago by grinding up some lean beef, freeze-drying it, and extracting every last bit of intramuscular fat with petroleum solvent. After removing any traces of the solvent, they rehydrated the meat and cooked up the patties, boiling them in plastic bags for greater standardization. The aroma of the burgers—surprisingly, after all this chemistry—was indistinguishable from ordinary beef. The missing fat just didn't matter. When they used chloroform and methanol to extract even the phospholipids, the resulting burgers had much less of the meaty aroma. So if you're a carnivore, thank the cell membranes for the meaty flavor of your next stew or steak.

This mix of fats differs subtly depending on which part of the animal the meat comes from, the breed of animal involved, and diet. Grain-fed beef, for example, has more monounsaturated fatty acids, including the very tasty oleic acid. Animals that eat a pasture diet, by contrast, end up with more polyunsaturated fats, plus a few additional flavor compounds such as skatole, a molecule that adds a pleasant funk at the concentration it's found at in meat, but which smells fecal at higher concentrations. But because cattle and sheep are ruminants—that is, they have complex stomachs where microbes break down their grassy diet, including the fats—the flavor of their meat doesn't depend strongly on diet. In contrast, pigs and chickens have simple stomachs and the fats in their diets more often make it into the meat intact. That's why you'll see specialty pork producers proudly touting animals finished on chestnuts or acorns—like the prized *jamón ibérico* of Spain—but rarely see specialized diets used as a selling point for beef.

Most of the flavor of meat develops when we start to cook it, as the heat of cooking begins to break down fatty acids into smaller molecules, many of which carry strong flavors. (Dry aging of meat also breaks down fatty acids, so aged meat develops even more of these flavors.) The weak point in a fatty acid molecule is the carbon-carbon double bonds, the "unsaturated" parts of the molecule, so polyunsaturated fatty acids have more weak points, and break into smaller molecules, than monounsaturated or saturated fats do. These fatty acid degradation products account for most of the meaty aromas and flavors in cooked meat. They're most obvious in meat that's cooked at relatively low temperatures, such as by simmering or stewing. At higher temperatures, when meat begins to brown, another process dominates the flavor picture.

The browning reaction—more formally known as the Maillard reaction after the early-twentieth-century French chemist who first described it—is responsible for a whole host of flavor changes that happen when foods cook. It's the reason why bread is much tastier after it's baked, why we roast our coffee beans, and why cauliflower roasted in the oven is more delicious than plain boiled cauliflower. It's the reason we grill our steaks instead of poaching them, and why the best stews start by browning the meat in a little fat.

Even though it's called *the* Maillard reaction, what we're really dealing with here is a vast network of interconnecting chemical reactions, almost like a braided streambed. At the upstream end, the reaction begins when amino acids and sugars react with each other to form a series of unstable intermediate compounds. Those intermediates then react with one another, and sometimes with fatty acids and other molecules in the vicinity as the process quickly becomes too complex to keep track of fully. These products give the characteristic brown color, and many of them are also volatile flavor molecules. Each food has its own unique starting point into the stream as a result of its particular mix of amino acids and sugars, so the reaction proceeds differently. That's why roast beef smells different from baking bread.

Lab chemists starting with pure amino acids and sugars have documented at least 621 different Maillard products; real foods, with their vastly greater diversity of chemical starting points, almost certainly produce even more products. We'll leave their detailed identification to the flavor chemists. For now, suffice it to say that Maillard products are responsible for all the toasty, roasty flavors we get in baked goods, roasted and grilled meats, and anything else with a browned crust. Most of these Maillard products are present in only tiny quantities, but our senses are exquisitely sensitive at detecting them. On the downside, the Maillard reaction can also produce molecules like acrylamide and other carcinogens. Chemists are hard at work trying to find ways to guide the reaction into stream courses that enhance the beneficial flavors and avoid these unhealthy molecules.

For the cook, the most important thing to know about the Maillard reaction is that it requires high temperatures, typically well above the boiling point of water. That's why fried and grilled foods brown, but stewed, steamed, and simmered foods don't. It's also why conscientious cooks dry the surface of their meat before searing—with less moisture to evaporate, the meat reaches Maillard temperatures more quickly so that more flavor develops. (Actually, the Maillard reaction does happen at lower temperatures, too—but so slowly that it rarely figures in cooking. Low-temperature Maillard reactions explain why powdered eggs sometimes turn brown after long storage, the impetus for some of the early research on the Maillard reaction. And black garlic, a cutting-edge ingredient these days, owes its complex, caramel-like flavor in part to Maillard reactions that take place over the span of a month at temperatures well below the boiling point.)

Because the Maillard reaction requires amino acids, or the proteins formed from them, it works most dramatically for protein-rich foods like meats, though most grains and vegetables also contain enough protein to generate some Maillard products. For some vegetables, especially sugar-rich ones like onions, a second browning reaction—caramelization—is also important. In caramelization, sugar molecules react with one another, rather than with amino acids, to form a similar cascade of flavorful products. Since sugars lack the nitrogen and sulfur atoms found in amino acids, however, caramelization produces a narrower range of flavor compounds, and less of the meaty, roasty flavors of Maillard products. From the cook's point of view, though, both can be treated as a single, high-temperature browning process.

As complex as browning is, cooks can still steer the process to some extent. Meat that contains more fat will feed more fatty acid breakdown products into the reaction, producing more of the roasty furans that make a rib roast of beef or a leg of lamb so delicious—a big reason we like to roast these cuts without trimming off all the surface fat. Cooking temperature makes a big difference, too, by pushing the flow down one branch or another of the Maillard stream.

For any carnivore, of course, all this raises a practical question: What's the best way to grill a steak? As it turns out, this has been the subject of sober scientific study by a meat scientist in (where else?) Texas named Chris Kerth. I phoned Kerth in his office at Texas A&M University, a hotbed of agricultural research, to get the scoop.

The hotter you cook a steak, the more you shift the balance from the beefy, brothy fatty acid breakdown products toward the roasty, nutty Maillard products. "You have a whole continuum that you can play with to customize the flavor," says Kerth. "There's a lot of restaurants where that's their claim to fame, is cooking their steaks at 1,800 degrees—which is obviously going to be for a very, very short period of time. That's what creates their signature flavor." For thicker steaks, the outside would burn black before the middle cooks adequately, so these restaurants often sear their steaks to develop the right Maillard crust, then finish cooking in a gentler oven.

Most of us don't have access to temperatures like that. To find out what works best at more typical cooking temperatures, Kerth embarked on the lab version of a cook-off. He bought whole beef strip loins and cut them into steaks either one-half, one, or one and one-half inches thick, then cooked the steaks to well done at one of three different temperatures: 350, 400, or 450 degrees Fahrenheit. It's hard to get the temperature exact on a grill, so Kerth cooked his steaks in preheated cast-iron skillets instead. (Science demands some sacrifices. The bigger sacrifice here, actually, is that Kerth fed his cooked steaks not to hungry Aggie volunteers, but to a gas chromatograph, again in the interest of greater precision.) As you'd expect, the thinner steaks (and those in the hotter pans) cooked through more quickly, which left less time for roasty Maillard flavors to develop. In thin steaks, as a result, tallowy, fatty, green flavors tended to dominate, while thicker steaks were more roasty, nutty, and buttery—but also had more acrid flavors. Since then, Kerth has also cooked experimental steaks for actual people, and he reports that most of them preferred the flavor of thick steaks cooked at relatively low temperatures. "That's been my recommendation, is to find a little bit lower temperature," he told me. "There's also an impact on tenderness—the lower, slower grilling results in more tender meat."

The third main way that cooks can create flavor in the kitchen is through fermentation, the process that creates such diverse flavor wonders as cheese, bread, soy sauce, kimchi, and beer and wine. Actually, fermentation is probably better described as a form of herding than cooking, because what we're really doing is managing the microbes that are doing the hard work of breaking down sugars and other molecules in the food, releasing volatile flavor molecules as they go. Often, a whole ecosystem of microbes—bacteria, yeasts, and other fungi—is involved in a fermentation. As we saw for wine in the previous chapter, the outcome of fermentation depends on exactly which microbes are involved.

That's easiest to see in the case of cheese. Several species of lactic acid bacteria attack the lactose in milk, generating sour-tasting lactic acid as a waste product. As the milk acidifies, its proteins curdle into a semisolid mass that the cheesemaker strains and presses to form the basic starting point for the cheese. Now things get far more diverse, as cheese makers can encourage different sets of microbes to take over the job. If a fungus called *Penicillium camemberti* settles in, its microscopic filaments form a whitish rind on the outside of the cheese and secrete enzymes that break down the casein protein, gradually liquifying the center of the cheese and generating the sharp, ammonia aromas of degraded proteins that mark a ripe Camembert. On the other hand, the related *Penicillium roqueforti* favors a different set of enzymes that break down the milk fats in the cheese, yielding sharp-flavored fatty acids and 2-heptanone, the signature flavor compound of blue cheeses such as Roquefort. Bacteria in Swiss cheese produce propionic acid, which contributes the nutty flavor of that cheese. The reddish rind of Limburger cheese is rich in the bacterium *Brevibacterium linens*, which produces sulfury by-products that give the cheese its stinky, body-odor quality (an apt analogy, since a related species lives in human armpits). Many other microbes contribute minor notes to the flavor of cheeses—indeed, the complexity these minor microbes add is the main reason for the deeper, more complex flavors of raw-milk cheeses. (These complex microbial ecosystems are also what makes sourdough bread more flavorful than bread risen from plain old, cultured baker's yeast.)

One of the big questions both professional chefs and amateur home cooks want to know about flavor is which ingredients go well together. Until recently, however, every cook—from the most primitive tribeswoman to world-famous chefs with three Michelin stars—has worked entirely by trial and error. We learn what goes well together by combining ingredients and seeing if the result tastes good. (Actually, most people cook what their culture has always cooked: Vietnamese savor fermented fish sauce, hot chilis, and lime; those from southern India favor mustard seed, coconut, and tamarind; southern Italians mix tomato, garlic, and basil. But this merely pushes the trial and error into the distant past.) The approach has obviously worked well, as any stroll through the restaurant districts of New York or San Francisco would reveal. But it's hard to get very far off the beaten path using trial and error. To really explore the far reaches of possibility, it would help a lot if we could find some general principles that underlie and guide our choice of flavor combinations that work.

You'll sometimes hear chefs cite the principle, "What grows together, goes together," as a basis for their flavor pairings. It's not hard to come up with some outstanding examples: morels with asparagus, lamb with thyme and rosemary from the Mediterranean hillsides where it once grazed, venison with cranberries and wild forest mushrooms. Zearfoss particularly likes the combination of apricots with the chanterelle mushrooms that grow in apricot orchards. But is there any scientific reason this principle should hold true?

At one level, yes—because it directs your attention to ingredients that are local and in season, and therefore most likely to be at their peak of flavor. Why would you not pair morels and asparagus in the springtime, when they're both at their best? At a slightly deeper level, the grows-together/goes-together principle can be seen as an endorsement of traditional flavor pairings. After all, for most of the history of civilization, cooks had no choice but to combine foods that grew together—local, seasonal food was all that was available (especially if you think of winter storage foods as another type of seasonal ingredient). Over the course of generations, cooks learned which combinations were most pleasant, and these became fixed by tradition. Meanwhile, no one much notices the pairings that didn't make the cut. (Spinach grows well in the springtime, too, but no one makes a big deal about pairing spinach with morels.) The upshot is that the grows-together/goes-together pairs that come to mind are largely the ones that have passed our ancestors' taste tests. Following one of those pairings is likely to yield a better result than you'd get with a random pairing of two ingredients—chili peppers and turnips, say—that aren't naturally found together and therefore haven't been vetted by tradition.

On the other hand, there's probably no basis in the science of flavor chemistry for expecting that ingredients from the same place would combine particularly well. As we've seen, the molecules that give fruits and vegetables their flavor do not come from the soil directly but are made by the plants themselves. That means there's no reason why two plants that grow together should be more likely to make similar flavor molecules, or molecules that are compatible in some other way.

We can push this a little further, though. Because our expectations and our previous experience play a big role in our perception of flavor, and especially in our flavor preferences, we might predict that familiar, traditional combinations of ingredients—based, of necessity, on foods that grow together—would strike us as more pleasant, on the whole, than novel combinations. What grows together, goes together not necessarily because it's intrinsically better, but because we've tried it before and we expect to like it.

There's another problem inherent in puzzling out pairs of ingredients that go well together: it can quickly become overwhelming, because the number of possible combinations explodes far too fast to evaluate all of them. Consider the pizza problem: If I have twenty-five different toppings, I can make twenty-five different one-topping pizzas, so it's reasonable to ask you which one you like best. But if I'm making a two-topping pizza, you've got six hundred different combinations to evaluate (that's 25 × 24, for the math geeks—we won't consider pepperoni and pepperoni as a two-topping option), and only an obsessive would work through the entire list to pick the best pair. And if you'd like a three-topping pie, you've got nearly fourteen thousand combinations to choose from. It's little wonder most pizzas use the same standard set of toppings, over and over and over again.

A few years ago, Michael Nestrud, a sensory scientist then at Cornell University who is also a CIA-trained chef, realized that an arcane branch of mathematics called graph theory might provide fresh solutions to the pizza problem by helping identify appealing food combinations more quickly. Despite its name, graph theory has nothing to do with the bar charts and zigzag lines most of us call graphs. Instead, it's all about groups of connected objects—in this case, foods that go together. Nestrud's insight was that you should be able to recognize a good three-topping pizza by seeing that each topping pairs well with the other two. Mathematically, this is identical to picking out "cliques" of Facebook friends where every member of the clique is friends with every other member.

So Nestrud made a list of pairs of possible pizza toppings and asked several hundred university students to give a thumbs-up or thumbs-down to each suggested pairing. From their answers, he compiled a list of "good" topping pairs, such as pepperoni and mushroom. Then he used the mathematics of graph theory to pull out sets of three or more toppings for which all the pairs were on the "good" list. These three-topping pizzas also turned out to be more popular than you'd expect by chance.

Of course, you don't need advanced mathematics to top a pizza. However, Nestrud's approach attracted serious interest from the U.S. Army, which desperately wanted to make tastier field rations. Soldiers in combat situations need food that is light, nutritious, and—above all—durable, and for decades, that has meant the dreaded MRE (the acronym stands for "meal, ready-to-eat"). MREs are precooked meals sealed into foil pouches. From the Army's point of view, MREs are great. They'll last for years, and soldiers can just grab them and go. The problem is, soldiers quickly get bored with the meals. It's already hard to get soldiers to eat enough when they could be shot or blown up at any moment, and boring food doesn't help matters. So the Army puts a lot of effort into making MREs as appealing as possible under the circumstances.

Each MRE contains an entree, side dishes, fruit, dessert, snacks, condiments, candy, and beverages, each chosen from up to thirty-two different options. These components could conceivably be mixed and matched in any which way—more than twenty-two billion different combinations in total. Which ones would the soldiers like? The Army hired Nestrud—fresh out of graduate school with his PhD in pizza toppings—to figure it out.

Using the same approach he took with the pizza, Nestrud designed a questionnaire that listed possible pairs of items and asked soldiers whether they'd want to eat them at the same meal: beef roast with vegetable couscous, meatballs and gravy with barbecue sauce, beef taco filling with jalapeño cheese spread, chicken fajita with bacon cheese spread, and so on. Using the pairs most commonly approved by the soldiers, Nestrud could then assemble entire MRE menus that he predicted would prove popular (top of the list: chili with beans, Mexican mac and cheese, herb-citrus seasoning, crackers with chunky peanut butter, fruit, cookies, and cheese pretzels), and others that he predicted the soldiers would hate. When he showed those menus to actual soldiers and asked them to rate their compatibility, the soldiers' ratings matched his predicted ones almost perfectly—real-world validation that you really can use Nestrud's approach to predict flavor pairings.

Nestrud's next move was to a consulting company, where he used his graph-theory technique to help identify which snack foods people like to buy at the same time. Grocery stores and fast-food restaurants could then display these "go togethers" next to each other in the store so that consumers who bought one might happen to buy the other, as well—seemingly on a whim, but really the result of careful thought and planning by the seller. (Nestrud has no idea whether his clients actually put his recommendations into practice.)

In his current job as sensory scientist with Ocean Spray, America's largest cranberry company, Nestrud is trying a different twist to this flavor-association game. Every day during the winter of 2015–2016, he searched through Twitter's daily archives and gathered every tweet that mentioned certain flavor-related key words. (The details are secret, of course, and Nestrud was careful not to mention the word "cranberry" when I spoke to him, but it's a pretty safe bet that it was one of his key words.) Cleaning up the data took a lot of work. He had to toss out key word hits that didn't really refer to flavor, such as references to cranberry-colored paint for the bathroom or mentions of orange that referred to the color of the uniforms of the Denver Broncos football team. And conversely, he had to ensure that "cranberry," "cran-apple," and "cran-raspberry" grouped together as similar flavors, and likewise "orange," "mandarin," and "tangerine."

In the first four months, Nestrud accumulated nearly twelve thousand relevant tweets—enough to get a good sense of what other flavors the Twitterverse thought of when it thought of cranberry. Better yet, his sample included both Thanksgiving and Christmas, as well as the postholiday lull, so he could see how flavor pairings changed through the seasons. The results might not lead directly to new products, but they're the first step in a long creative process. "The ultimate goal is not to make any final decisions," says Nestrud. "It's to generate hypotheses about products we wouldn't have thought of on our own that we can then go out and validate with real consumer testing."

Professional chefs like to push the limits of tradition, and so do adventurous eaters. One of the great joys of eating is to discover a novel mix of ingredients that works unexpectedly well together, breaking us out of the comfortable confines of tradition and into a new world of possibility. We can find these new combinations through trial and error, of course, or we can rely on the intuition of gifted cooks, which is essentially trial and error inside the cook's imagination. But perhaps our search for delicious novelty can also get some guidance from what we know about the science of flavor.

One of the first really promising steps in this direction came a decade ago, when world-famous chef Heston Blumenthal of the Fat Duck restaurant was experimenting with salty ingredients in desserts, and discovered that white chocolate and caviar make a great flavor combination. The pairing was so bizarre—yet so delicious—that Blumenthal mentioned it to a colleague at an industrial flavor company. A little work soon revealed that both members of this unlikely pairing are rich in a compound called trimethylamine, which has a fishy flavor.

This got Blumenthal thinking. If a shared flavor molecule

accounts for the success of this odd pairing, maybe similar "molecular rhymes" might point us to other surprising flavor combinations. The idea makes some intuitive sense. As we've seen, chefs often balance similarity and contrast in building their dishes—and since flavor is all about molecules, then similar flavors should share flavor molecules. As Blumenthal pursued these molecular similarities, he came up with a whole kitchenful of brilliant, unexpected matches: liver and jasmine, which share sulfur compounds; carrots and violets, which share a molecule called ionone; pineapple and blue cheese; snails and beets.

In the years that followed, Blumenthal's insight sparked a whole gastronomic movement. Going by the name of "food pairing," it makes these molecular rhymes the centerpiece of combining foods. There's even a commercial service (foodpairing .com) that, for the price of a monthly subscription, will let professional chefs and enthusiastic amateurs start with any ingredient and follow a web of molecular similarities to find other foods with supposedly complementary flavors.

A Canadian sommelier named François Chartier has begun to investigate the pairing of wine with food along the same lines, based on chemical similarities between ingredients in the food and aroma compounds in the wine. For instance, Chartier suggests pairing a rosemary-scented lamb stew with a dry Riesling wine, to take advantage of the citrusy, floral-smelling molecules in that wine, which echo those in the rosemary. This "molecular sommellerie" was novel enough to earn Chartier's book on the subject, *Taste Buds and Molecules*, a prize as the "best innovative food book in the world" at the 2010 Gourmand World Cookbook Awards.

You'd think, with all this excitement, that food scientists would be eager to sink their teeth into molecular food pairing to see if it really works. But very few have actually done so—and even fewer have published their results in the scientific literature. (Food-pairing, Inc., the company selling food-pairing ideas to chefs, hasn't released any evidence to back up its approach.)

The obvious test would be to have people rate how well pairs of ingredients go together, and see whether those that share more flavor molecules receive higher ratings. Wender Bredie, a Danish food scientist at the University of Copenhagen, did exactly that a few years ago, using fifty-three different pairs of ingre-dients ranging from cinnamon and apple to cinnamon and garlic, malt and cocoa to malt and blue cheese. They found that the number of flavor molecules in common made absolutely no difference to the rated pleasantness. "I have never seen an experimental study with a correlation that was so low," recalls Bredie. (It's worth noting that Bredie's study—like an earlier study by a different group that reached much the same result—was only presented at a scientific conference, not published in a scientific journal. That means the research has not been vetted by other experts, so the conclusions should be regarded as preliminary.)

Bredie's group did find one interesting result, though. Pairs with fewer molecules in common tended to be perceived as more novel than those that shared more molecules—a feature at least a few high-end chefs might like to make use of, though novelty is not the same as pleasantness. "With high-end restaurants, you want something unique and surprising for your customers," says Bredie. "And it doesn't have to be nice. If you go to Noma"—the New Nordic restaurant in Copenhagen that was rated number one in the world for several years—"the foods aren't very nice. You go there, and you have a fantastic experience. But if you ask the customers, 'Is this something you'd really like to eat more frequently?' they'd probably say no." (When Bredie says that Noma's dishes "aren't very nice," he means that the ingredients and techniques are often unusual and challenging, such as the moss cooked in chocolate that is featured on the menu as I write this.)

A second approach would be to look at the flavor combinations that people actually use, and count the molecules their ingredients have in common. If these real combinations are more likely to share flavor molecules than random sets of ingredients are, that would be evidence that molecular rhymes really do make combinations taste better. The data are out there: the Internet age has provided a huge treasure trove of real flavor combinations, in the form of online recipes, and anyone with a few hundred dollars to spare can subscribe to a database that lists all the flavor compounds in any given food ingredient. The big challenge is making sense of the tangled web of recipes, ingredients, and flavor molecules.

Enter Sebastian Ahnert. By day a theoretical physicist at the University of Cambridge and by night an enthusiastic amateur cook, Ahnert has exactly the skill set necessary to pick apart the problem. A few years ago, he and his colleagues downloaded more than fifty-six thousand recipes from three online recipe archives (Epicurious, Allrecipes, and a Korean database called Menupan) and studied their molecular overlaps. Real recipes, they found, had a slight tendency to share more flavor molecules in common than random collections of ingredients—but only for North American, Western European, and Latin American cuisines, where common ingredients like milk, eggs, butter, and wheat share overlapping flavor profiles. Asian recipes actually shared fewer molecules than random ingredients, because their most common ingredients, such as soy sauce, scallion, ginger, and rice, have largely nonoverlapping flavors. When Ahnert left those most common ingredients out of his analysis, he found no evidence at all to support the food-pairing hypothesis.

Ahnert's analysis made a big splash when it was published in a topnotch scientific journal, but he wasn't satisfied. Recipes aren't an ideal starting point, because some ingredients—flour and eggs come to mind—are often included more for structural reasons than because they contribute important flavors. So Ahnert went back to the drawing board. This time, instead of recipes, he used ingredient pairings recommended by well-known chefs, which he found in a best-selling book called *The Flavor Bible* by Karen Page and Andrew Dornenburg. He found that chef-recommended pairs share more flavor molecules than randomly paired ingredients—and the pattern gets stronger when he counts only the most abundant flavor molecules, or those with the most food-related odors.

So maybe there's something to food-pairing theory after all. Not everyone is convinced yet, particularly because as I write this, Ahnert has not yet published his most recent reanalysis. But even if foods that go well together do tend to share flavor molecules, that's not the same as saying that foods that share flavor molecules necessarily go well together. This molecular-rhyming approach is probably an idea generator at best.

To explore a really high-tech approach to finding unusual, exciting flavor combinations, I visited IBM's Thomas J. Watson Research Center in Yorktown Heights, New York. The company has a long history of taking on the biggest challenges in artificial intelligence, and the Watson Research Center is where it all happens. IBM's Deep Blue supercomputer made headlines back in 1997 when it beat world chess champion Garry Kasparov in a six-game match. Then, in 2011, Deep Blue's successor, Watson (named, like the research center itself, after the company's long-time president from the first half of the twentieth century), beat two human champions at the quiz game *Jeopardy*. With Watson's win under their belts, IBM's researchers started looking for new applications for their expertise. Why not, they asked, turn Watson's immense powers to the kitchen? After all, cooking is both highly creative and a familiar, everyday activity that millions of people do regularly. And Watson's computing prowess should pay off big time, because the computer could learn more recipes, more ingredients, more techniques than any human ever could, just like it had for *Jeopardy* trivia. Let's do it, the Watson team decided.

The Thomas J. Watson Research Center sits in forested hills just off the Taconic State Parkway, less than an hour's drive north of midtown Manhattan. The main building—a vast, three-story curving facade, designed by renowned architect Eero Saarinen—looms over visitor parking, and the main entrance, sheltered by a thrusting, flaring overhang, leads into a 1960s-futuristic foyer. It's all very high concept, very expensive looking, and very formal. Just what you'd expect from IBM, the corporation long notorious for its severely conservative dress code.

In such a setting, Florian Pinel comes as a distinct shock. What you first notice about the French-born software engineer is not his broad face, blue eyes, or unruly mop of stringy brown hair. It's the four stainless-steel studs that pierce the corners of his mouth and lower lip, and the longer, blade-shaped fin that emerges from the center of his lower lip, just above his chin. Dressed in jeans and a ratty shirt, instead of IBM's traditional white shirt and tie,

Pinel looks more at home in a restaurant kitchen than in an IBM conference room.

Appearances aren't deceiving in this case—Pinel is indeed comfortable in the kitchen. While working at IBM, he spent his weekends studying at New York's respected Institute of Culinary Education, earning a chef's ticket in 2005. For a while after that, he worked Saturday nights as a line cook in a Manhattan restaurant, just for the thrill of it. "That was a big rush," he recalls. Eventually, though, he gave that up and focused on cooking at home in his newfound leisure time. When Watson came along, he was ready.

How do you teach a computer how to cook? Not the way you'd teach your kid, by having him or her stand by your elbow and watch. And not the way Pinel learned, in culinary school. Instead, you feed the computer data. Lots and lots of data. Food chemists have identified the key flavor chemicals in most ingredients, and psychologists have measured how pleasing we find each of them. Cyberspace is full of recipes that show how people all over the world cook: which ingredients they use, and how they combine them. Pinel and his team input all this information into Chef Watson's memory bank. From this mass of data, the computer chef extracted patterns: the sets of ingredients that were likely to go well together, and the sequence of steps to use in combining those ingredients. (A big help for the latter was getting access to *Bon Appétit* magazine's archive of nine-thousand-plus recipes, all tested and carefully edited into a standard format.)

Anyone with a computer can consult Chef Watson for free (at least as of this writing) at ibmchefwatson.com. You simply type in an ingredient or two, and Chef Watson suggests a few other ingredients you might try. Once you've settled on a core set of four ingredients—and, optionally, specified a style such as French, summer, or vegetarian—you can choose from a list of suggested recipes, complete with measurements and cooking techniques. It's that simple.

Behind the scenes, though, a lot is going on. To come up with its recommendations for ingredients that work together, Chef Watson looks for ingredients that are already used together in existing recipes somewhere in the world, or sets of ingredients that share several flavor chemicals like Heston Blumenthal's white chocolate and caviar. But for Pinel, finding these sets of ingredients alone isn't enough to make Watson truly creative. "We think there are two things that make something creative," he says. "It has to be novel, and it has to be valuable." For a recipe, "value" equals deliciousness—something that Chef Watson could estimate from its knowledge of which flavor chemicals people like best, and from its calculations of chemical overlaps. And novelty was an easy one—Watson just calculated how similar a recipe's ingredients are to those used in other recipes. Tomatoes, garlic, oregano? Not very novel. Asparagus, pig's feet, and Indian spices? You bet. For each ingredient combination, Chef Watson gives you a "synergy" score—essentially a composite of compatibility, pleasantness, and surprise. A high synergy score means Watson is confident of its choice of ingredients, says Pinel. "This is going to work well, and it's not going to be trivial, either."

For a foodie, software like this is endlessly intriguing, and it's easy to get sucked down the rabbit hole, browsing one idea after another. But it doesn't take much experimentation to realize that for all of Chef Watson's knowledge and computing power, it lacks the finely honed kitchen instincts of a Michelin-starred chef. Instead, it's more like your brilliant-but-loopy buddy who blurts out whatever thought happens to pass through his mind, no matter how bizarre. I tested Chef Watson in late January, right around Robbie Burns Day, the great Scottish haggis-and-whisky fest. Since the traditional accompaniment, "neeps and tatties" (turnips and potatoes, for the uninitiated), is traditionally boring, I thought I'd see whether Chef Watson had any better ideas. I started with the key ingredient, turnips, and specified that I wanted a Scottish recipe. Then—to continue the Scottish theme—I added another of that country's favorite foods: beer.

The chef's suggestion: "Scottish Turnip Meatball," a veal/turkey meatball served in a sauce of chili powder, the Indian spice mix called garam masala, turnip, avocado, and clam juice. It sounds like a bizarre jumble, and I almost didn't try it. But in the spirit of research, I finally subjected my family to it one night—and, to our surprise, it worked pretty well. The creamy unctuousness of the avocado made just the right counterpoint to the turnip's bitterness, and the garam masala and clam juice added a subtle depth to the flavor profile. In fact, it was good enough that we served it to dinner guests a few weeks later. Maybe there really is something to this whole molecular food-pairing thing.

Want something to cook for your Super Bowl party? Just specify "Super Bowl" in the "Pick a Style" field, and Chef Watson suggests some ingredients to start with: raisins, garlic, chocolate chips, and endive. The synergy score is off the charts, well over 90 percent, so for some reason Chef thinks this is a great combination. I think I'll try another spin of the wheel.

Watson's next suggestion: pork belly, shallot, ginger, and white pepper. Ah, more promising. One of the suggested recipes for that combination is Superbowl Pork Belly Bolognese, which sounds plausible. But the recipe itself is over the top, calling for ground chicken breast, ground pork belly, and ground chicken wings (bones in, I wonder?), as well as chorizo sausage and a quarter cup of horseradish.

To drink? Why not a Cauliflower Bloody Mary, made with Pernod and ouzo, not vodka or gin, and swapping out the seasoned tomato juice for a puree of cauliflower, shiitake mushrooms, and onion. Garnish with grape wedges, Chef suggests—a peculiarly fussy take on the usual lime wedge, and certainly one I wouldn't have thought of on my own. (Weird recipe steps like that show up frequently, because Watson bases its procedures on existing *Bon Appétit* recipes, substituting similar ingredients as needed. In this case, probably, Watson decided to substitute grapes for lime because both are fruits with plenty of acid, and it just borrowed the technique verbatim. That likely also explains the ground chicken wings in the pork belly bolognese. You'll soon find your own humorous combinations.)

I'm poking fun here, because Chef Watson leaves some low-hanging fruit. But even its goofiest ideas often have a kernel of real inspiration. When I started with Italian sausage and broccoli, it suggested a recipe adapted from a braised brisket dish: rub the sausage with a seasoning mix, "working the paste into all the cracks," then put the sausage in a casserole "fat side up." Clearly Watson doesn't understand the difference between a brisket and a sausage. But that night as I lay in bed, I realized that a spice rub might add a nice touch to something as prosaic as a grilled bratwurst, or even a hot dog. Good idea, Chef. And maybe, with the right drink, a garnish of grape wedges wouldn't be such a bad idea after all.

The jury's still out on whether Chef Watson is a major step forward in culinary creativity or just an amusing sideline. So far, the app is generating about fifty thousand ingredient pairings a month, says Pinel—which sounds like a lot until I realize that I've probably done fifty today, all by myself. Some users just look for suggestions of ingredient pairs, and build or adapt their own recipes; others click on complete recipes. The next step, says Pinel, is to add nutritional information to the mix, so that Chef Watson can double as Dietician Watson.

With menu chosen and kitchen chemistry properly deployed to marshal just the right flavor molecules, there's one more step that a savvy cook can take to boost the flavor of a meal: serve it properly. We've seen that presentation can legitimately be considered as part of the flavor experience: Changing the color, shape, or weight of the plate or bowl can make the food taste sweeter or more bitter. Charles Spence, the psychologist behind that study, took that notion still further in another experiment. Working with Charles Michel, a professional chef, Spence presented volunteers with a salad containing identical ingredients plated in one of three ways. Some diners got an ordinary tossed salad, some got a salad with each ingredient stacked neatly in separate piles, and some got a salad that was dramatically plated in a splash of colors and shapes that resembled a Wassily Kandinsky painting. The diners who got the Kandinsky salad found it both more pleasing aesthetically and also tastier than those who got the boring versions. For any cook, at home or in a restaurant, an attractive presentation is more than just window dressing—it makes the food itself more flavorful.

We can apply the same principle to wine: drinking it from an elegant glass should make it taste better. There are functional reasons for this as well as the psychological ones. A large, tulip-shaped glass that tapers inward toward the lip offers more space for volatile molecules to gather above the liquid, enhancing the aroma. Studies have verified that the same wine does indeed taste more pleasant from a glass like this than from a straight-sided water glass. On the other hand, there's little hard evidence that you get any extra flavor boost from having a different glass shape for Bordeaux-style wines than for Burgundies, as some high-end crystal makers suggest. I recommend spending your money on the wine, instead. (I asked one expert in the oenology department at the University of California, Davis, who's done some of the wine glass studies, what kind of wine glasses she uses. "Whatever I get free from the winery," she replied, with a laugh.)

Many wine lovers like to decant their wine before serving, especially for reds. Besides the obvious benefit of eliminating the sediment that some wines accumulate in the bottle, decanting is supposed to "let the wine breathe" and improve its flavor. In molecular terms, decanting allows the escape of some of the off flavors that may have developed in the bottle, and it may also allow oxygen—nearly absent in the bottle—to react with the wine and create some new flavor compounds. Whatever the reason, it does seem to work.

But if a little decanting is good, would more be better? Nathan Myhrvold thinks so. Myhrvold, the former chief technology officer at Microsoft (and before that a physicist who studied under Stephen Hawking) has, in recent years, taken an engineer's approach to high-end cuisine, where he clearly enjoys finding iconoclastic approaches to familiar kitchen tasks. For wine, Myhrvold recommends "hyperdecanting" by pouring the wine into a blender and buzzing it on high for thirty to sixty seconds. "Even legendary wines, like the 1982 Chateau Margaux, benefit from a quick run through the blender," he writes in his six-volume cookbook *Modernist Cuisine*.

Of course I had to try it (though not with the Chateau Margaux, which is well beyond my means). I decanted one-third of a bottle of wine the usual way, by pouring it into a decanter, and put a third into the blender, keeping the last third undecanted in the bottle. Then I had my son, who was underage at the time, pour all three into numbered glasses so that the rest of the dinner party could taste them without knowing which was which. Myhrvold turned out to be right—sort of. The wine from the blender practically leaped out of the glass, with a huge and vivid aroma. It was much better than the other two, which we all thought indistinguishable. But when I returned to the glass five or ten minutes later, the blenderized wine was lifeless and spent—it appeared to have no flavor left. If I'm pouring six or eight small glasses from a bottle to be consumed immediately, I'd definitely go the blender route. If I'm going to share a bottle with my wife over a leisurely dinner, though, I'll keep the blender in the cupboard and pour the normal way.

Epilogue

THE FUTURE OF FLAVOR

The science of flavor is burgeoning. Every month, researchers publish new studies about our flavor senses, the psychology and neuroscience of flavor perception, and techniques to enhance flavor in industrial food labs, on the farm, and in the kitchen. We know more about flavor than ever before, and new vistas are opening all the time.

That bright outlook, full of promise for the future, is true for flavor's standing in our everyday lives, as well. Case in point: Recently, my wife and I spent a few days driving through the mountains of eastern British Columbia, Canada. It's a pretty remote place, a good day's drive from the major cities of Vancouver, Calgary, and Edmonton. Yet almost everywhere we went—in towns of ten thousand to forty thousand people—we drank excellent local craft beer. The only exception, a town of just four thousand, had big beer plans, too, but we were a few months too early for their microbrewery's opening.

Nor is eastern British Columbia an exception here. The number of breweries in the United States has risen from a nadir of fewer than ninety in 1978 to more than four thousand today, and almost all of them are small craft brewers. The boom continues, with the number of craft breweries increasing by nearly 20 percent every year. After decades of mostly boring, mass-market beers, anyone interested in distinctive, flavorful beer suddenly has more options than they can keep track of. Big breweries have taken note and are introducing their own "craft beers" under different labels. Even the United Kingdom, which has long treasured its local brews, is in the midst of an expansion, with the number of breweries more than doubling since 2000.

You can see this flavor renaissance in other foods, too. Just walk down the aisles of any midsized grocery store and look at all the flavors offered that our parents and grandparents never knew. Sriracha and other hot sauces share the shelves with ketchup. Instead of just rice, we have basmati, jasmine, and arborio, not to mention red rice, black rice, sticky rice, and more. The produce section has habanero chilis, fennel, and arugula—and often dragon fruit, bitter melon, and curry leaves as well. The spice shelves offer not just parsley, sage, rosemary, and thyme, but also star anise, garam masala, and smoked paprika. If you're willing to step outside the convenience of the supermarket, you'll find an even wider range of tasty choices at ethnic markets, greengrocers, and the plethora of farmers' markets that have sprung up in the last decade or two.

Eating out tonight? Especially if you're in a large city—but often even if you're not—you can choose among sushi bars, noodle shops, Thai or Vietnamese restaurants, northern or southern Indian places. Feel like Chinese? Take your pick of Cantonese, Szechuan, Beijing, Shanghai, or Hunan cuisines, plus Fukien, Hakka, or some other regional specialty if you're lucky. You can find food from the Middle East, from Northern Africa, and from Mexico, Spain, Italy, and France. With a little effort, you can eat Afghan, Russian, Brazilian, or Peruvian food. Truly, this is the golden age of flavor.

It wasn't always this way, of course. In the middle of the twentieth century, much of the English-speaking world was mired in a dark age of flavor. Readers who are old enough may recall such dishes as canned green beans topped with canned cream-of-mushroom soup, JELL-O salad with canned fruit cocktail and nondairy whipped topping, and TV dinners. Restaurant options midcentury usually included little more than roadside diners, generic Chinese, or stuffy French or "Continental" restaurants, followed by a tidal wave of fast-food burgers and french fries in the 1950s and 1960s.

The culprit in much of this was a push for efficiency, in the name of modernization. Early in the twentieth century, the domestic science movement sought to base household management—including cooking—on scientific principles, delivering calories and nutrients with a minimum of fuss and effort. The movement brought such horrors as Crisco white sauce: Crisco shortening, flour, and milk, cooked up into a flavorless paste of empty calories. As refrigerated transport became more common, growers began to select varieties that shipped well and looked good in the market, rather than the most flavorful ones. This preference gave us iceberg lettuce, Red Delicious apples, and, as we've seen, the supermarket tomato.

Beginning in the 1950s, advertising by the growing food-processing industry began pushing the notion that their convenient products were the solution to kitchen drudgery. Home cooks increasingly turned to canned soups, TV dinners, and other shortcuts, with help from books like *The Can-Opener Cookbook*. When even that was too much work, fast food was always an option. As more and more families had both parents working, these options became increasingly attractive—and flavor ended up the loser in the deal.

But even as the culinary mainstream turned blander, the first seeds of today's flavor renaissance were beginning to germinate. Soldiers returning from the Second World War brought home their experiences of new, foreign, delicious foods, and the rise of international air travel in the 1950s and 1960s helped more people expand their flavor worlds. Starting in the 1960s, liberalized immigration laws opened borders to a flood of new, non-European arrivals, who brought their traditional foods with them and became part of an expanding food culture. Consider, as one example, the fact that we take for granted today that almost everyone can eat with chopsticks, a skill that was the height of multicultural sophistication not too many decades ago.

The nascent food culture slowly gathered momentum. Beginning in the late 1960s, the counterculture movement emphasized healthy, homemade foods (albeit often earnestly stodgy ones). Pioneering restaurants like Chez Panisse in the San Francisco Bay Area (and counterparts elsewhere in the United States, as well as in Britain and Australia) focused attention on fresh, high-quality ingredients—where possible, locally grown, and including unfamiliar, rich flavors such as wild mushrooms and fresh herbs. Both restaurant chefs and home cooks began to seek out farmers' markets for fresh fruits and vegetables, often finding a wider range of more flavorful varieties than those available in the grocery store. The number of farmers' markets in America rose from just one hundred in 1960 to more than eight thousand in 2014, and new markets open every year.

Cookbooks, too, began to reflect a growing interest in food. Instead of "boring basics" tomes, cookbook buyers sought out more adventurous guides to more exciting, flavorful food. Julia Child's *Mastering the Art of French Cooking*, first published in 1961, is of course the most famous example, but my shelves are full of others, too: Time-Life's Foods of the World series, published between 1968 and 1972, Diana Kennedy's *The Cuisines of Mexico* (1972), Marcella Hazan's *The Classic Italian Cook Book* (1973, and its several successors), and Julee Rosso and Sheila Lukins's *The Silver Palate Cookbook* (1979). More recently, the Food Network and other televised outlets have turned food and flavor into a competitive sport, with chefs competing to make the most delicious and innovative dishes right in front of the cameras.

The Slow Food movement is also helping to bring flavor to the fore. Founded in 1986 by Italian journalist Carlo Petrini in reaction to the opening of a McDonald's in Rome, Slow Food now claims over 100,000 members in 1,500 local groups around the world. One of their key initiatives is the Ark of Taste, an effort to identify and preserve local food traditions that are in danger of disappearing. As of this writing, for example, the Ark's 3,277 entries include 321 from the United States—everything from the "old-type" Rhode Island Red chicken (bred for eating as well as laying, unlike the modern laying-only variety) to the yellow watermelons grown by the Tohono O'odham people of southern Arizona. The United Kingdom has 98 entries, including artisanal

Cheddar from Somerset, the Gloucester Old Spot pig, and Wessex Einkorn wheat. Every one of the Ark's entries represents a unique flavor experience—and thanks to the spotlight Slow Food is shining on them, they are experiences that are now a little less likely to vanish.

Flavor has never been more important in our culture, and it's a fair bet that its star will continue to rise in the future. (Why would anyone who learns to appreciate flavor ever turn the clock back?) But exactly where we're going is anyone's guess. Flavor preferences alter over time as the culture changes. If I served you a meal from medieval England today, you'd probably find its extravagant sweetness and overbearing use of cinnamon and cloves just as strange as Chaucerian English. A few generations from now, the way we cook will no doubt seem as peculiar, in retrospect, to our own descendants.

In the short term, though, we can feel confident about a few predictions. As societies become more multicultural, we'll see increasing fusion of culinary traditions. When I visited the Culinary Institute of America in Hyde Park, New York, for example, I had coffee with culinary anthropologist Willa Zhen in the Apple Pie Bakery Café, a classic American breakfast/lunch spot with pastries, sandwiches, soups, and salads. In addition to her anthropological schooling, Zhen also trained as a traditional Chinese chef, so she's well attuned to cross-cultural influences. She pointed out one of the day's specials, chicken noodle soup. Sounds pretty standard—but this one was made with Japanese udon noodles, Chinese cabbage, cilantro, and scallions. "This place is called the Apple Pie Café, and it's supposed to be as American as apple pie," she said, "and yet, we have Asian chicken noodle soup on the menu."

We can also predict that as our population ages, cooks will have to tinker with their seasonings to compensate for our declining flavor senses. Some obvious possibilities include using more salt and MSG, and perhaps more spicy-hot peppers to stimulate our dulling palates. But the changes may be more precise than that. Bob Sobel, vice president of research at the flavor company FONA, says they're working to understand whether certain classes of flavor molecules get harder to detect than others. If they should find that aging makes us more sensitive to sulfury thiols but less sensitive to fruity esters, for example, then elder-directed seasonings might play up the esters and back off the thiols. "It's not turning the volume up, it's adjusting the graphic equalizer," notes Sobel.

Finally, we can expect to see new flavors enter our repertoires as new ingredients take the stage. There's already a buzz going about insects as a tasty, inexpensive, sustainable protein source, and as the world's population grows, we're likely to see more insects on our menus. That may not sound appetizing to most of us today, but it's worth remembering that the tomato was viewed with at least as much suspicion when it first came to Renaissance Italy from the New World. Look where that ended up. If insects—which are already eaten with gusto in places like Mexico and Thailand—move into the mainstream, I'll bet they take their first steps through pairing with familiar seasoning flavors that help ease the shock of the new.

Wherever we're going, almost all of us can learn to get more from our everyday flavor experiences. Most of us never pay much attention to developing our flavor skills, so we make do with a vague and fuzzy flavor sense. Sure, we can tell a good chocolate cake from a bad one, or we might recognize that the peach we're eating in July is much more flavorful than the one we had back in January. But we can deepen that experience by recognizing that the July peach carries hints of coconut aromas, and that its intense sweetness was balanced by greater acidity, less astringency, and a juicier texture.

At this point you might protest that you don't have what it takes to pick out coconut notes in your peach or a barnyard aroma in your glass of wine. It's easy to think that the people who can do that sort of thing are blessed with exceptional palates that the rest of us can't hope to match.

But if there's one lesson I'd like you to take away from this book, it's that almost anyone can get better at appreciating flavor. It doesn't matter if you have trouble articulating precisely what you're tasting. As I've noted, we're all terrible at putting names to flavors unless we have prompts to fall back on. But if you're aware that one glass of wine tastes different from another, or that a Gala apple tastes different from a Red Delicious, or a raspberry from a strawberry, then don't worry. You've got the basic sensory abilities. All the rest is practice and attention.

Some cultures already do better at this than most of the English-speaking world does. For French schoolchildren, lunch is a subject like any other, where they learn how to appreciate traditional foods properly. The Slow Food movement aims to bring similar "taste education" to other cultures, including our own. Such training, they hope, can give anyone a better appreciation of flavor.

Even the pros don't necessarily start out with exceptional palates. Prominent wine critics don't tend to volunteer to have their sensory abilities tested. Who, after all, would take the risk that they might be branded a substandard taster? The little bit of information that is available, however, suggests that they're nothing special. Researchers in New Zealand, for example, measured olfactory thresholds of eleven wine professionals—winemakers, wine sellers, wine judges, and even wine researchers—and eleven ordinary people, and found no difference between the two groups. (Wine experts are slightly more likely than ordinary folks to report the intense bitter perceptions that mark supertasters, but it's not clear why that's a professional advantage.) High-fidelity wine tasters, in other words, are made, not born.

The same is true for professional flavorists. "It's a matter of learning and passion," one longtime professional flavorist told me about his trade. "I wouldn't say I'm a top 1 percent taster. I don't think that's the most important criterion. You don't have to be in the top 1 percent to be successful." That's good news for all of us ordinary amateurs who'd like to improve their flavor perception.

If you're one of that group, the way forward is simply to begin. Try this: Next time you eat an apple, don't just munch as you read this book or check your e-mail. Concentrate, instead, on your flavor experience. Give it your full attention. Try to articulate what you're tasting. How sweet is the apple? How tart? Do you get any bitterness from the skin? Is it rich or poor in that fruity, appley fragrance? Finally—and most important—ask yourself how well you like this apple. You may even find it helps to assign numbers to your sensations: score each quality from zero to ten, say. There's no better way to crystallize your perceptions than to force yourself to quantify them.

It will seem a little strange at first, all this pondering and scoring. You might feel self-conscious and a little pretentious, and you'll almost certainly struggle to put your flavor experience into words, even inside your own head. That's how I felt. But it gets easier with practice, and soon you'll find you're noticing subtler shades of sweetness, or comparing the aromatic fruitiness of a Macintosh to the gentle sweetness of a Fuji. After a while, you may find that you notice subtler flavor notes: a hint of banana in one, a touch of pear in another.

You can apply the same analytic skills you're developing with the apples to any other food. Whatever you're eating, slow down and pay close attention to the balance of flavors. See if you can pick out the herbs and spices in the stew, and whether the cook browned the onions before adding the liquid. Even if you've just grabbed a Big Mac on the run, try to pause for a moment and savor it. Many highly trained flavorists labored long and hard over that special sauce, and someone decided exactly how much sesame seed should be on that bun. See if you agree with their choices.

I don't eat like this all the time, of course. Sometimes I forget. Other times I'm distracted and wolf down my meal without really noticing, just like I usually did before. But I'm trying to pay attention and eat mindfully more often, and I'm slowly building my flavor chops. The more I do it, the easier it is to identify subtler elements of my flavor experience, and the more I build my flavor vocabulary, the better to describe what I'm tasting.

For professional flavorists, of course, paying attention has become second nature. When I visited Givaudan, the world's largest flavor company, several of the people I talked to noted that most of their flavorists stop and sniff everything before they put it in their mouths. Occasionally, the habit makes for a slightly awkward moment at a dinner party. "People ask me, 'Is something wrong with the food?'" one flavorist told me a bit sheepishly. (Maybe, in light of that, you want to think twice about flavor experiments in certain social settings.)

Practice is exactly how people come to grips with the complex flavors of wine, too. You can learn the basic dimensions of wine description—color, body, astringency, acidity, sweetness—from any number of books. For the more subtle elements of the flavor—the hint of anise, or blackberry, or tobacco—feel free to play around. Buy some cheap red wine and divide it into a half-dozen jars. Mash a few raspberries in one, a slice of plum in the next, a few blackberries in another, and so on, to make aroma standards. Then pick a jar at random and see whether you can identify the addition just by smelling the wine. (The guy at my local wine shop was amused when I asked him to recommend the least flavorful, most nondescript wine on his shelves for this exercise.) This is exactly the method that the wine mavens at UC Davis use when training tasting panels to recognize wine aromas.

It also helps immensely to fall back on a flavor wheel or other crib sheet. These days you can find flavor wheels online for everything from wine, beer, and Scotch whisky to cheese, chocolate, and coffee. (I found one for apples, too!) Look around to see what you can find for some of your own favorite foods. Having a list of potential flavors to choose from avoids the tip-of-your-tongue problem, when you know what a flavor is but can't call up a name.

I now carry in my wallet a folded wine-aroma crib sheet that I can pull out when I'm at a loss for words (although I do temper my geekiness when I'm with friends). Just last week, for example, I found an odd but familiar aroma in a glass of California Zinfandel. I couldn't immediately name it, but I recognized it as soon as I saw it on my crib sheet: horse sweat (it tasted much better than it sounds, actually). I was surprised to identify it; I'd never tasted that in a wine before, but the crib sheet made me confident in my call.

Of course, my confidence could be misplaced, but I try not to let that hold me back. Remember, even expert perfumers and flavorists can't accurately identify more than three or four aromas from a mixture. In something as complex as wine, that means the experts' flavor identifications miss the mark pretty often. (You can easily verify that by comparing two critics' reviews of the same wine and noting their lack of overlap.) The bottom line is that accuracy doesn't matter. What's important is that coming up with a description forces me to pay attention, and paying attention enriches my flavor experience. It slows me down, so that meals become a time for dining, not merely for eating.

There's a world of flavor out there waiting, and it's ours to enjoy.

ACKNOWLEDGMENTS

I had no idea a book required so much cooperation from so many people. More than one hundred scientists and flavor professionals generously gave their time and knowledge to educate me about flavor. Only a few of them are mentioned by name here, but every one, named or not, helped shape my understanding in ways great and small. Thank you all—it's been fascinating, every step of the way.

A few deserve special mention. Leslie Stein arranged my visit to the Monell Chemical Senses Center, where Joel Mainland and Dani Reed sequenced portions of my genome and ran me through a panel of perceptual tests, and Gary Beauchamp knocked out my sense of taste. Linda Bartoshuk in Florida examined my tongue and shared her vast knowledge of taste. Richard Doty tested my sense of smell and let me visit his taste disorders clinic at the University of Pennsylvania for a day, and Patricia Yager graciously let me write about her case. Andreas Keller gave up most of one Saturday to show me around his lab and talk about smell. I spent three fascinating days with Bob Sobel and the other flavorists at FONA International, and another day with Brian Mullin and the flavorists at Givaudan. Profound thanks to Tracy Cesario at FONA and Jeff Peppet at Givaudan for setting up the visits. Nicole Gaudette at Alberta Agriculture arranged my participation in a consumer taste panel. Maynard Kolskog at the Northern Alberta Institute of Technology let me invade his research kitchen and cooked strange food with me. Chris Loss and Jonathan Zearfoss talked flavor over a lovely meal at the CIA. Thanks to all of you for letting me participate.

For phone conversations above and beyond the normal interview, I'm grateful to Sanne Boesveldt, John Hayes, Michael Nestrud, Charles Spence, Dana Small, Mike Trought, Carol Wagstaff, and Vance Whittaker.

Sebastian Ahnert, Bruce Bryant, Tracy Cesario, Richard Doty, Harry Klee, Darren Logan, Joel Mainland, Richard Mattes, Florian Pinel, Charles Spence, Leslie Stein, Mike Trought, Carol Wagstaff, Vance Whitaker, Patricia Yager, and Jonathan Zearfoss valiantly read whole chapters or portions thereof to help me get the facts straight. Any errors that remain are my own fault.

Thanks to Richard Doty for permission to reprint portions of the University of Pennsylvania Smell Identification Test, and to Caroline Hobkinson for graciously allowing me to reproduce portions of the menu from her Multisensory Feast at the House of Wolf restaurant.

I'm indebted to Justin Mullins for suggesting that there was a book in this, and to my agent, Peter Tallack at The Science Factory, for encouraging me to develop the proposal and then finding just the right home for it. Louisa Pritchard has done a wonderful job selling foreign rights.

I thank my editor, John Glusman at W. W. Norton, for his steady support and sensitive editing—and for a well-timed kick in the posterior, without which I'd still be interviewing chefs and flavor scientists! Thanks, too, to my British editors, Ed Faulkner and Elen Jones at Ebury Press. Alexa Pugh and Lydia Brents at Norton patiently answered all my rookie questions, and kept the whole thing moving in the right direction. Rebecca Homiski and Louise Mattarelliano shepherded the manuscript through the production process with care. Nina Hnatov's meticulous copyediting was a joy to behold. I thought I had an eye for detail, but Nina showed me a whole new level. Susan Groarke's proofreading ensured that the text made the trip from manuscript to proof intact and error-free. I'm immensely grateful to Chin-Yee Lai for a stunning cover design and Chris Welch for an elegant layout. I'm grateful to Elizabeth Riley, Meredith McGinnis, and Steve Colca in Norton's marketing department, and to Bill Rusin and his sales team for guiding this book's journey to your hand.

I couldn't have done this without personal support. Joel Shurkin, David Quammen, Ed Struzik, and John Acorn shared their book-writing advice. Gordon Fox and Kathy Whitley housed me in Florida, and Mark and Lisa Holmes did the same in New Jersey. My colleagues at *New Scientist* magazine tolerated my periodic disappearances. Many friends helped keep me sane, but Ed Struzik, Jim and Karin Stewart, Alan Nursall, Heidi Zwickel, and Mike Sullivan deserve special mention. My neighbors, especially Tony and Wanda next door, have tolerated years of indifferent lawn care.

Finally, and most important, I thank my family. My parents, John and Kathleen Holmes, exposed me to lots of unusual flavors as a kid, and have encouraged me ever since. Most of all, I thank my wife, Deb Moon, and our son Ben for all their love and support, and for being willing guinea pigs for lots of flavor experiments.

NOTES

INTRODUCTION

4 *Richard Wrangham argues*: Richard Wrangham, *Catching Fire: How Cooking Made Us Human* (New York: Basic Books, 2009): 105–127.

4 *spices have antibacterial properties*: Paul W. Sherman and Jennifer Billing, "Darwinian Gastronomy: Why We Use Spices," *BioScience* 49 (1999): 453–463.

5 *more brain systems*: Gordon M. Shepherd, "Neuroenology: How the Brain Creates the Taste of Wine," *Flavour* 4 (2015): 19, doi:10.1186/s13411-014-0030-9.

6 *Vegemite*: Paul Rozin and Michael Siegal, "Vegemite as a Marker of National Identity," *Gastronomica* 3, no. 4 (2003): 63–67.

6 *holidays, sex, and family time*: J. Westenhoefer and V. Pudel, "Pleasure from Food: Importance for Food Choice and Consequences of Deliberate Restriction," *Appetite* 20 (1993): 246.

8 *English speakers generally use taste*: Paul Rozin, "'Taste-Smell Confusions' and the Duality of the Olfactory Sense," *Perception and Psychophysics* 31 (1982): 397–401.

11 *Bush told reporters*: Maureen Dowd, "'I'm President,' So No More Broccoli!," *New York Times*, March 23, 1990, http://www.nytimes.com/1990/03/23/us/i-m-president-so-no-more-broccoli.html.

CHAPTER 1: BROCCOLI AND TONIC

14 *Bartoshuk who first suggested*: Linda M. Bartoshuk, Valerie B. Duffy, and Inglis J. Miller, "PTC/PROP Tasting: Anatomy, Psychophysics, and Sex Effects," *Physiology & Behavior* 56 (1994): 1165–1171.

17 *lose those extraneous tastes*: Alexander A. Bachmanov et al., "Genetics of Taste Receptors," *Current Pharmaceutical Design* 20 (2014): 2669–2683.

17 *vampire bats*: Wei Hong and Huabin Zhao, "Vampire Bats Exhibit Evolutionary Reduction of Bitter Taste Receptor Genes Common to Other Bats," *Proceedings of the Royal Society B* (2014), doi:10.1098/rspb.2014.1079.

23 *104 diverse bitter-tasting chemicals*: Wolfgang Meyerhof et al., "The Molecular Receptive Ranges of Human TAS2R Bitter Taste Receptors," *Chemical Senses* 35 (2010): 157–170.

26 *published a letter*: Robert Ho Man Kwok, "Chinese-Restaurant Syndrome," *New England Journal of Medicine* 278 (1968): 796.

26 *picked up the story*: For a good treatment of the history of Chinese restaurant syndrome, see Ian Mosby, "'That Won-Ton Soup Headache': The Chinese Restaurant Syndrome, MSG and the Making of American Food, 1968–1980," *Social History of Medicine* (2009): 133–151, doi:10.1093/shm/hkn098.

27 *fifty-eight million pounds*: Ibid., 7.

27 *most damning evidence*: L. Tarasoff and M. F. Kelly, "Monosodium L-Glutamate: A Double-Blind Study and Review," *Food and Chemical Toxicology* 31 (1993): 1019–1035.

27 *when researchers looked back*: Ibid.

29 *discovered by accident*: Anonymous, "The Inventor of Saccharine," *Scientific American*, July 17, 1886, 36.

29 *Cyclamate*: Deborah Jean Warner, *Sweet Stuff: An American History of Sweeteners from Sugar to Sucralose* (Lanham, MD: Rowman & Littlefield, 2011), 195, accessed via Google Books, March 29, 2016.

29 *Aspartame*: Robert H. Mazur, "Discovery of Aspartame," in Lewis D. Stegink and L. J. Filer, Jr., eds., *Aspartame: Physiology and Biochemistry* (New York: Marcel Dekker, 1984), 4.

29 *Sucralose*: Burkhard Bilger, "The Search for Sweet," *The New Yorker*, May 22, 2006, 40.

30 *Pepsi is about 11 percent*: Daniel Engber, "The Quest for a Natural Sugar Substitute," *New York Times Magazine*, January 1, 2014, http://www.nytimes.com/2014/01/05/magazine/the-quest-for-a-natural-sugar-substitute.html.

30 *its own distinctive timing*: Paul A. S. Breslin and Alan C. Spector, "Mammalian Taste Perception," *Current Biology* 18 (2008): R153.

30 *ten seconds later*: Engber, "Quest for a Natural Sugar."

30 *four seconds longer*: Ibid.

31 *9 grams of salt daily*: S. L. Drake and M. A. Drake, "Comparison of Salty Taste and Time Intensity of Sea and Land Salts from around the World," *Journal of Sensory Studies* 26 (2010): 25.

31 *from processed foods*: Marjorie Ellin Doyle and Kathleen A. Glass, "Sodium Reduction and Its Effect on Food Safety, Food Quality, and Human Health," *Comprehensive Reviews in Food Science and Food Safety* 9 (2010): 44–56.

31 *high blood pressure*: Ibid., 45.

32 *they taste saltier*: Tassyana Vieira Marques Freire et al., "Salting Potency and Time-Intensity Profile of Microparticulated Sodium Chloride in Shoestring Potatoes," *Journal of Sensory Studies* 30 (2015): 1–9.

33 *bitter-tasting medicines*: Adam A. Clark, Stephen B. Liggett, and Steven D. Munger, "Extraoral Bitter Taste Receptors as Mediators of Off-Target Drug Effects," *FASEB Journal* 26 (2012): 4827–4831.

34 *more sinus infections*: Robert J. Lee and Noam A. Cohen, "The Emerging Role of the Bitter Taste Receptor T2R38 in Upper Respiratory Infection and Chronic Rhinosinusitis," *American Journal of Rhinology and Allergy* 27 (2013): 283–286.

35 *receptors on our taste buds*: Robin M. Tucker, Richard D. Mattes, and Cordelia A. Running, "Mechanisms and Effects of 'Fat Taste' in Humans," *BioFactors* 40 (2014): 313–326.

35 *a distinct taste*: Cordelia A. Running, Bruce A. Craig, and Richard D. Mattes, "Oleogustus: The Unique Taste of Fat," *Chemical Senses* 40 (2015), 507–516.

35 oleogustus: Ibid.

36 *a taste for calcium*: Michael G. Tordoff et al., "T1R3: A Human Calcium Taste Receptor," *Scientific Reports* 2 (2012): 496, doi:10.1038/srep00496.

36 *for carbon dioxide*: Jayaram Chandrashekar et al., "The Taste of Carbonation," *Science* 326 (2009): 443–445.

36 *a taste for starch*: Breslin and Spector, "Mammalian Taste Perception," R149.

37 *calcium-sensing receptor*: Motonaka Kuroda and Naohiro Miyamura, "Mechanism of the Perception of 'Kokumi' Substances and the Sensory Characteristics of the 'Kokumi' Peptide, Gamma-Glu-Val-Gly," *Flavour* 4 (2015): 11, doi:10.1186/2044-7248-4-11.

37 *interact with one another*: Russell S. J. Keast and Paul A. S. Breslin, "An Overview of Binary Taste-Taste Interactions," *Food Quality and Preference* 14 (2002): 117.

37 *ability to taste PROP*: Bernd Bufe et al., "The Molecular Basis of Individual Differences in Phenylthiocarbamide and Propylthiouracil Bitterness Perception," *Current Biology* 15 (2005): 322–327.

38 *That's probably why*: Bartoshuk, Duffy, and Miller, "PTC/PROP Tasting."

39 *support that hunch*: For example, John E. Hayes and Valerie B. Duffy, "Revisiting Sugar-Fat Mixtures: Sweetness and Creaminess Vary with Phenotypic Markers of Oral Sensation," *Chemical Senses* 32 (2007): 225–236.

39 *fail to find a link*: For example, Mary E. Fischer et al., "Factors Related to Fungiform Papillae Density: The Beaver Dam Offspring Study," *Chemical Senses* 38 (2013): 669–677; Nicole L. Garneau et al., "Crowdsourcing Taste Research: Genetic and Phenotypic Predictors of Bitter Taste Perception as a Model," *Frontiers in Integrative Neuroscience* 8 (2014): 33, doi:10.3389/fnint.2014.00033.

39 *gustin might be involved*: Melania Melis et al., "The Gustin (CA6) Gene Polymorphism, rs2274333 (A/G) as a Mechanistic Link between PROP Tasting and Fungiform Taste Papilla Density and Maintenance," *PLoS One* 8 (2013): e74151, doi:10.1371/journal.pone.0074151.

40 *affects sweet perception*: Alexey A. Fushan et al., "Allelic Polymorphism within the TAS1R3 Promoter Is Associated with Human Taste Sensitivity to Sucrose," *Current Biology* 19 (2009): 1288–1293.

42 *two kinds of supertasters*: Natalia V. Ullrich et al., "PROP Taster Status and Self-Perceived Food Adventurousness Influence Food Preferences," *Journal of the American Dietetic Association* 104 (2004): 543–549.

CHAPTER 2: BEER FROM THE BOTTLE

48 *unique pattern of molecular vibrations*: For the most detailed exposition of this point of view, see Luca Turin, *The Secret of Scent: Adventures in Perfume and the Science of Smell* (London: Faber and Faber, 2006).

49 *a different chord*: This apt analogy is not mine, alas. I first encountered it on the Food Sommelier website, http://www.foodsommelier.com/sensory_reality/.

49 *used it in their key paper*: Linda Buck and Richard Axel, "A Novel Multigene Family May Encode Odorant Receptors: A Molecular Basis for Odor Recognition," *Cell* 65 (1991): 183.

50 *But a closer look*: This paragraph follows Avery Gilbert, *What the Nose Knows: The Science of Scent in Everyday Life* (New York: Crown, 2008): 2–4.

53 *413 odor receptors*: Tsviya Olender et al., "Personal Receptor Repertoires: Olfaction as a Model," *BMC Genomics* 13 (2012): 414, doi:10.1186/1471-2164-13-414.

55 *ORs all over the place*: Ester Feldmesser et al., "Widespread Ectopic Expression of Olfactory Receptor Genes," *BMC Genomics* 7 (2006): 121, doi:10.1186/1471-2164-7-121.

55 *combine ethyl isobutyrate*: E. Le Berre et al., "Just Noticeable Differences in Component Concentrations Modify the Odor Quality of a Blending Mixture," *Chemical Senses* 33 (2008), 389–395.

55 *one part geraniumy*: C. Masanetz, H. Guth, and W. Grosch, "Fishy and Hay-like Off-flavours of Dry Spinach," *Zeitschrift für Lebensmitteluntersuchung und -Forschung A* 206 (1998): 108–113.

56 *a trillion different odor objects*: C. Bushdid et al., "Humans Can Discriminate More Than 1 Trillion Olfactory Stimuli," *Science* 243 (2014): 1370–1372.

56 *treated with caution*: Richard C. Gerkin and Jason B. Castro, "The Number of Olfactory Stimuli That Humans Can Discriminate Is Still Unknown," *eLife* 4 (2015): e08127, doi:10.7554/eLife.08127.

57 *a terrific book*: Gordon M. Shepherd, *Neurogastronomy* (New York: Columbia University Press, 2012).

57 *asked her to close her eyes*: Yaara Yeshurun and Noam Sobel, "An Odor Is Not Worth a Thousand Words: From Multidimensional Odors to Unidimensional Odor Objects," *Annual Review of Psychology* 61 (2010): 226.

58 *straight to a neurologist*: The colleague is Jay Gottfried of Northwestern University. See Greg Miller, "What's Up with That: Why Are Smells So Difficult to Describe in Words?" *Wired*, November 11, 2014, http://www.wired.com/2014/11/whats-up-with-that-smells-language/.

59 *"Big Red gum"*: This quote is from Asifa Majid and Niclas Burenhult, "Odors Are Expressible in Language, as Long as You Speak the Right Language," *Cognition* 130 (2014): 266–270.

59 *quick and consistent*: Ibid.

60 *wine experts' noses are no better*: Wendy V. Parr, David Heatherbell, and K. Geoffrey White, "Demystifying Wine Expertise: Olfactory Threshold, Perceptual Skill and Semantic Memory in Expert and Novice Wine Judges," *Chemical Senses* 27 (2002): 747–755.

60 *fell off dramatically*: D. G. Laing and G. W. Francis, "The Capacity of Humans to Identify Odors in Mixtures," *Physiology & Behavior* 46 (1989): 809–814.

60 *Later studies have confirmed*: Anthony Jinks and David G. Laing, "A Limit in the Processing of Components in Odour Mixtures," *Perception* 28 (1999): 395–404.

61 *ambrosial and stench*: Stanley Finger, *Origins of Neuroscience: A History of Explorations into Brain Function* (Oxford, UK: Oxford University Press, 2001), 178.

61 *The Suya*: Constance Classen, David Howes, and Anthony Synnott, *Aroma: The Cultural History of Smell* (London: Routledge, 1994), 100–101.

62 *The Serer-Ndut*: Ibid., 102–104.

64 *chocolate-tracking experiment*: Jess Porter et al., "Mechanisms of Scent-Tracking in Humans," *Nature Neuroscience* 10 (2007): 27–29.

66 *don't get any better*: Lee Sela and Noam Sobel, "Human Olfaction: A Constant State of Change-Blindness," *Experimental Brain Research* 205 (2010): 13–29.

66 *smell your hand*: Idan Frumin et al., "A Social Chemosignaling Function for Human Handshaking," *eLife* 4 (2015): e05154, doi:10.7554/eLife.05154.

66 *Sobel told a reporter*: Catherine de Lange, "After Handshakes, We Sniff People's Scent on Our Hand," *New Scientist*, March 3, 2015, https://www.newscientist.com/article/dn27070-after-handshakes-we-sniff-peoples-scent-on-our-hand/.

67 *one famous experiment*: Daniel J. Simons and Daniel T. Levin, "Failure to Detect Changes to People During a Real-World Interaction," *Psychonomic Bulletin and Review* 5 (1998): 644–649.

67 *change blindness*: Sela and Sobel, "Human Olfaction."

68 *forms an air curtain*: Rui Ni et al., "Optimal Directional Volatile Transport in Retronasal Olfaction," *Proceedings of the National Academy of Sciences* 112 (2015): 14700–14704.

68 *according to Shepherd*: Shepherd, *Neurogastronomy*, 19–27.

70 *thresholds tend to be lower*: Viola Bojanowski and Thomas Hummel, "Retronasal Perception of Odors," *Physiology & Behavior* 107 (2012): 484–487.

71 *smells different to each nostril*: Noam Sobel et al., "The World Smells Different to Each Nostril," *Nature* 402 (1999): 35.

72 *30 percent of our odor receptors*: Joel D. Mainland et al., "The Missense of Smell: Functional Variability in the Human Odorant Receptor Repertoire," *Nature Neuroscience* 17 (2014): 114–120.

74 *genes help determine*: Charles J. Wysocki and Gary K. Beauchamp, "Ability to Smell Androstenone Is Genetically Determined," *Proceedings of the National Academy of Sciences* 81 (1984), 4899–4902.

76 *can vary many thousandfold*: Andreas Keller et al., "An Olfactory Demography of a Diverse Metropolitan Population," *BMC Neuroscience* 13 (2012): 122, doi:10.1186/1471-2202-13-122.

76 *nothing special*: For example, Parr, Heatherbell, and White, "Demystifying Wine Expertise."

77 *easier to detect*: David E. Hornung et al., "Effect of Nasal Dilators on Nasal Structures, Sniffing Strategies, and Olfactory Ability," *Rhinology* 39 (2001): 84–87.

77 *More than a thousand other genes*: Ifat Keydar et al., "General Olfactory Sensitivity Database (GOSdb): Candidate Genes and Their Genomic Variations," *Human Mutation* 34 (2012): 32–41.

77 *"transforms my chamber pot"*: Quoted in Marcia Levin Pelchat et al., "Excretion and Perception of a Characteristic Odor in Urine after Asparagus Ingestion: A Psychophysical and Genetic Study," *Chemical Senses* 36 (2011): 9–17.

77 *a pound of canned asparagus*: M. Lison, S. H. Blondheim, and R. N. Melmed, "A Polymorphism of the Ability to Smell Urinary Metabolites of Asparagus," *British Medical Journal* 281 (1980): 20–27.

78 OR2M7: Nicholas Eriksson et al., "Web-Based, Participant-Driven Studies Yield Novel Genetic Associations for Common Traits," *PLoS Genetics* 6 (2010): e1000993, doi:10.1371/journal.pgen.1000993.

78 *who really do produce odorless urine*: Marcia Pelchat et al., "Excretion and Perception of a Characteristic Odor in Urine after Asparagus Ingestion: A Psychophysical and Genetic Study," *Chemical Senses* 36 (2010): 9–17.

78 OR6A2 *gene*: Nicholas Eriksson et al., "A Genetic Variant near Olfactory Receptor Genes Influences Cilantro Preference," *Flavour* 1 (2012): 22, doi:10.1186/2044-7248-1-22.

81 *Smell-O-Vision*: See Gilbert, *What the Nose Knows*: 155–163.

83 *recent German study*: Andreas Dunkel et al., "Nature's Chemical Signatures in Human Olfaction: A Foodborne Perspective for Future Biotechnology," *Angewandte Reviews* 53 (2014): 7124–7143.

CHAPTER 3: THE PURSUIT OF PAIN

87 *the receptor for capsaicin*: Michael J. Caterina et al., "The Capsaicin Receptor: A Heat-Activated Ion Channel in the Pain Pathway," *Nature* 389 (1997): 816–824.

88 *extra-virgin olive oil*: Catherine Peyrot des Gachons et al., "Unusual Pungency from Extra-Virgin Olive Oil Is Attributable to Restricted Spatial Expression of the Receptor of Oleocanthal," *Journal of Neuroscience* 31 (2011): 999–1009.

95 *report less burn*: Pamela Dalton and Nadia Byrnes, "The Psychology of Chemesthesis: Why Would Anyone Want to Be in Pain?," in Shane T. McDonald, David Bolliet, and John Hayes, eds., *Chemesthesis: Chemical Touch in Food and Eating* (Chichester, UK: Wiley, 2016), 8–31.

96 *58 percent of our liking*: Outi Tornwall et al., "Why Do Some Like It Hot? Genetic and Environmental Contributions to the Pleasantness of Oral Pungency," *Physiology & Behavior* 107 (2012): 381–389.

96 *"benign masochism"*: Paul Rozin and Deborah Schiller, "The Nature and Acquisition of a Preference for Chili Pepper by Humans," *Motivation and Emotion* 4 (1980): 77–101.

98 *more likely to be sensation seekers*: Nadia K. Byrnes and John E. Hayes, "Personality Factors Predict Spicy Food Liking and Intake," *Food Quality and Preference* 28 (2013): 213–221.

98 *an interesting pattern emerged*: Nadia K. Byrnes and John E. Hayes, "Gender Differences in the Influence of Personality Traits on Spicy Food Liking and Intake," *Food Quality and Preference* 42 (2015): 12–19.

99 *fifty-hertz vibration*: Nobuhiro Hagura, Harry Barber, and Patrick Haggard, "Food Vibrations: Asian Spice Sets Lips Trembling," *Proceedings of the Royal Society B* 280 (2013): 1680, doi:10.1098/rspb.2013.1680.

100 *blocks the flow of potassium*: Kristin A. Gerhold and Diana M. Bautista, "Molecular and Cellular Mechanisms of Trigeminal Chemosensation," *Annals of the New York Academy of Sciences* 1170 (2009): 184–189.

100 *high-altitude climber*: Mark Graber and Stephen Kelleher, "Side Effects of Acetazolamide: The Champagne Blues," *American Journal of Medicine* 84 (1988): 979–980.

101 *eliminating the bubbles*: Paul M. Wise et al., "The Influence of Bubbles on the Perception Carbonation Bite," *PLoS One* 8 (2013): e71488, doi:10.1371/journal.pone.0071488.

103 *astringency built up*: Catherine Peyrot des Gachons et al., "Opponency

of Astringent and Fat Sensations," *Current Biology* 22 (2012): R829–R830.

103 *reported the first hints*: Nicole Schöbel et al., "Astringency Is a Trigeminal Sensation That Involves the Activation of G Protein-Coupled Signaling by Phenolic Compounds," *Chemical Senses* 39 (2014): 471–487.

CHAPTER 4: THIS IS YOUR BRAIN ON WINE

109 *sugar tasted sweeter*: Richard J. Stevenson, John Prescott, and Robert A. Boakes, "Confusing Tastes and Smells: How Odours Can Influence the Perception of Sweet and Sour Tastes," *Chemical Senses* 24 (1999): 627–635.

109 *host of similar studies*: For an overview of these, see Malika Auvray and Charles Spence, "The Multisensory Perception of Flavor," *Consciousness and Cognition* 17 (2008): 1016–1031.

111 *Pringles potato chips*: Massimiliano Zampini and Charles Spence, "The Role of Auditory Cues in Modulating the Perceived Crispness and Staleness of Potato Chips," *Journal of Sensory Studies* 19 (2004): 347–363.

111 *sounds of a coffee maker*: Klemens Michael Knöferle, "Acoustic Influences on Consumer Behavior: Empirical Studies on the Effects of In-Store Music and Product Sound," (PhD dissertation, University of St. Gallen, 2011), 36, http://www1.unisg.ch/www/edis.nsf/SysLkpByIdentifier/3964/$FILE/dis3964.pdf.

112 *when accompanied by the sea sounds*: Charles Spence, Maya U. Shankar, and Heston Blumenthal, "'Sound Bites': Auditory Contributions to the Perception and Consumption of Food and Drink," in Francesca Bacci and David Melcher, eds., *Art and the Senses* (Oxford, UK: Oxford University Press, 2011), 225–226.

113 *words like* kiki: Alberto Gallace, Erica Boschin, and Charles Spence, "On the Taste of 'Bouba' and 'Kiki': An Exploration of Word-Food Associations in Neurologically Normal Participants," *Cognitive Neuroscience* 2 (2011): 34–46.

113 *ice cream called "Frosh"*: Eric Yorkston and Geeta Menon, "A Sound Idea: Phonetic Effects of Brand Names on Consumer Judgments," *Journal of Consumer Research* 31 (2004): 43–51.

114 *the crockery*: This aspect is discussed extensively in Charles Spence and Betina Piqueras-Fiszman, *The Perfect Meal: The Multisensory Science of Food and Dining* (Chichester, UK: Wiley, 2014): 109–143.

115 *sweetness, but not its saltiness*: J. A. Maga, "Influence of Color on Taste Thresholds," *Chemical Senses* 1 (1974): 115–119.

115 *had tinted it red*: Gil Morrot, Frédéric Brochet, and Denis Dubourdieu, "The Color of Odors," *Brain and Language* 79 (2001): 309–320.

116 *three different rooms*: Carlos Velasco et al., "Assessing the Influence of the Multisensory Environment on the Whisky Drinking Experience," *Flavour* 2 (2013): 23, doi:10.1186/2044-7248-2-23.

116 *ice hockey games*: Corinna Noel and Robin Dando, "The Effect of Emotional State on Taste Perception," *Appetite* 95 (2015): 89–95.

119 *squirm-inducing experiment*: Dana M. Small et al., "Differential Neural Responses Evoked by Orthonasal versus Retronasal Odorant Perception in Humans," *Neuron* 47 (2005): 593–605.

121 *Shepherd puts it best*: Gordon M. Shepherd, *Neurogastronomy* (New York: Columbia University Press, 2012), ix (emphasis in original).

123 *labeled as "body odor"*: Ivan E. de Araujo et al., "Cognitive Modulation of Olfactory Processing," *Neuron* 46 (2005): 671–679.

125 *results were shocking*: Robert T. Hodgson, "An Examination of Judge Reliability at a Major U.S. Wine Competition," *Journal of Wine Economics* 3 (2008): 105–113.

125 *other major wine competitions*: Robert T. Hodgson, "An Analysis of the Concordance Among 13 U.S. Wine Competitions," *Journal of Wine Economics* 4 (2009): 1–9.

126 *not humanly possible to judge wines objectively*: I owe this idea to Anna Katharine Mansfield, a wine researcher at Cornell University.

126 *prefer cheaper wines*: Robin Goldstein et al., "Do More Expensive Wines Taste Better? Evidence from a Large Sample of Blind Tastings," *Journal of Wine Economics* 3 (2008): 1–9.

127 *wines of varying price*: Hilke Plassmann et al., "Marketing Actions Can Modulate Neural Representations of Experienced Pleasantness," *Proceedings of the National Academy of Sciences* 105 (2008): 1050–1054.

128 *The frontal operculum*: Janina Seubert et al., "Superadditive Opercular Activation to Food Flavor Is Mediated by Enhanced Temporal and Limbic Coupling," *Human Brain Mapping* 36 (2015): 1662–1676.

129 *in monkey brains*: Edmund T. Rolls et al., "Sensory-Specific Satiety: Food-Specific Reduction in Responsiveness of Ventral Forebrain Neurons after Feeding in the Monkey," *Brain Research* 368 (1986): 79–86.

129 *gradually switch their responses*: Edmund T. Rolls et al., "Orbitofrontal Cortex Neurons: Role in Olfactory and Visual Association Learning," *Journal of Neurophysiology* 75 (1996): 1970–1981.

131 *smell more accurately*: Jahan B. Jadauji et al., "Modulation of Olfactory Perception by Visual Cortex Stimulation," *Journal of Neuroscience* 32 (2012): 3095–3100.

CHAPTER 5: FEEDING YOUR HUNGER

134 *Sclafani offered rats*: Catalina Pérez, François Lucas, and Anthony Sclafani, "Increased Flavor Acceptance and Preference Conditioned by the Postingestive Actions of Glucose," *Physiology & Behavior* 64 (1998): 483–492.

136 *learned which flavor delivered*: Ivan E. de Araujo et al., "Metabolic Regulation of Brain Response to Food Cues," *Current Biology* 23 (2013): 878–883.

137 *An old study from the 1950s*: James Olds and Peter Milner, "Positive Reinforcement Produced by Electrical Stimulation of Septal Area and Other Regions of Rat Brain," *Journal of Comparative and Physiological Psychology* 47 (1954): 419–427.

138 *conscious and unconscious valuations*: Deborah W. Tang, Lesley K. Fellows, and Alain Dagher, "Behavioral and Neural Valuation of Foods Is Driven by Implicit Knowledge of Caloric Content," *Psychological Science* 25 (2014): 2168–2176.

141 *a glass of carrot juice*: Julie A. Mennella, Coren P. Jagnow, and Gary K. Beauchamp, "Prenatal and Postnatal Flavor Learning by Human Infants," *Pediatrics* 107 (2001): E88, http://www.pediatrics.org/cgi/content/full/107/6/e88.

142 *as infants liked it better*: R. Haller et al., "The Influence of Early Experience with Vanillin on Food Preference Later in Life," *Chemical Senses* 24 (1999): 465–467.

142 *more accepting of vegetable flavors*: Julie A. Mennella, "Ontogeny of Taste Preferences: Basic Biology and Implications for Health," *American Journal of Clinical Nutrition* 99 (2014): 704S–711S.

143 *"you are turning green"*: Carol Zane Jolles, *Faith, Food, and Family in a Yupik Whaling Community* (Seattle: University of Washington Press, 2002), 284; cited in Sveta Yamin-Pasternak et al., "The Rotten Renaissance in the Bering Strait: Loving, Loathing, and Washing the Smell of Foods with a (Re)acquired Taste," *Current Anthropology* 55 (2014): 619–646.

144 *wearing latex gloves*: Yamin-Pasternak et al., "Rotten Renaissance."

144 *didn't respond the same way*: Paul M. Wise et al., "Reduced Dietary Intake of Simple Sugars Alters Perceived Sweet Taste Intensity but Not Perceived Pleasantness," *American Journal of Clinical Nutrition* 103 (2016): 50–60.

146 *the renowned French Laundry*: Thomas Keller, *The French Laundry Cookbook* (New York: Artisan, 1999): 14.

147 *tomato soup aroma*: Mariëlle Ramaekers et al., "Aroma Exposure Time and Aroma Concentration in Relation to Satiation," *British Journal of Nutrition* 111(2014): 554–562.

147 *some other young Dutch men*: Anne G. M. Wijlens et al., "Effects of Oral and Gastric Stimulation on Appetite and Energy Intake," *Obesity* 20 (2012): 2226–2232.

148 *big squirts separated*: Dieuwerke P. Bolhuis et al., "Both Longer Oral Sensory Exposure to and Higher Intensity of Saltiness Decrease Ad Libitum Food Intake in Healthy Normal-Weight Men," *Journal of Nutrition* 141 (2011): 2242–2248.

149 *eating pasta with a small spoon*: Ana M. Andrade et al., "Does Eating Slowly Influence Appetite and Energy Intake When Water Intake Is Controlled?" *International Journal of Behavioral Nutrition and Physical Activity* 9 (2012): 135, doi:10.1186/1479-5868-9-135.

149 *playing with texture*: K. McCrickerd and C. G. Forde, "Sensory Influences on Food Intake Control: Moving beyond Palatability," *Obesity Reviews* 17 (2015): 18–29.

149 *soup more filling*: Mieke J. I. Martens and Margriet S. Westerterp-Plantenga, "Mode of Consumption Plays a Role in Alleviating Hunger and Thirst," *Obesity* 20 (2012): 517–524.

150 *highly flavored vanilla custard*: René A. de Wijk et al., "Food Aroma Affects Bite Size," *Flavour* 1 (2012): 3, doi:10.1186/2044-7248-1-3.

150 *saltier tomato soup*: Bolhuis et al., "Longer Oral Sensory Exposure."

150 *a dozen rat-friendly flavors*: Michael Naim et al., "Energy Intake, Weight Gain, and Fat Deposition in Rats Fed Flavored, Nutritionally Controlled Diets in a Multichoice ('Cafeteria') Design," *Journal of Nutrition* 115 (1985): 1447–1458.

151 *scarf the stuff down anyway*: Israel Ramirez, "Influence of Experience on Response to Bitter Taste," *Physiology & Behavior* 49 (1991): 387–391.

152 *any taste receptor or odor receptor genes*: Adam E. Locke et al., "Genetic Studies of Body Mass Index Yield New Insights for Obesity Biology," *Nature* 518 (2015): 197–206.

156 *tells you nothing useful*: Chih-Hung Shu et al., "The Proportion of Self-Rated Olfactory Dysfunction Does Not Change across the Life Span," *American Journal of Rhinology & Allergy* 23 (2009): 413–416.

156 *impairments of their sense of smell*: Claire Murphy et al., "Prevalence of Olfactory Impairment in Older Adults," *JAMA* 288 (2002): 2307–2312.

156 *scratch-and-sniff smell survey*: Charles J. Wysocki and Avery N. Gilbert, "National Geographic Smell Survey: Effects of Age Are Heterogeneous," *Annals of the New York Academy of Sciences* 561 (1989): 12–28.

158 *responded to both*: Nancy E. Rawson et al., "Age-Associated Loss of Selectivity in Human Olfactory Sensory Neurons," *Neurobiology of Aging* 33 (2012): 1913–1919.

158 *four times as likely to die*: Jayant M. Pinto et al., "Olfactory Dysfunction Predicts 5-Year Mortality in Older Adults," *PLoS One* 9 (2014): e107541, doi:10.1371/journal.pone.0107541.

159 *depression and anxiety*: Carl M. Philpott and Duncan Boak, "The Impact of Olfactory Disorder in the United Kingdom," *Chemical Senses* 39 (2014): 711–718.

159 *linked to other health problems*: Nicole Toussaint et al., "Loss of Olfactory Function and Nutritional Status in Vital Older Adults and Geriatric Patients," *Chemical Senses* 40 (2015): 197–203.

160 *might improve with practice*: Thomas Hummel et al., "Effects of Olfactory Training in Patients with Olfactory Loss," *Laryngoscope* 119 (2009): 496–499.

161 *Mark Friedman thinks*: David S. Ludwig and Mark I. Friedman, "Increasing Adiposity: Cause or Consequence of Overeating?" *Journal of the American Medical Association* 311 (2014): 2167–2168.

162 *Dana Small thinks*: Martin G. Myers Jr. et al.,"Obesity and Leptin Resistance: Distinguishing Cause from Effect," *Trends in Endocrinology and Metabolism* 21 (2010): 643–651.

162 *containers of sugary or fatty food*: Michael G. Tordoff, "Obesity by Choice: The Powerful Influence of Nutrient Availability on Nutrient Intake," *American Journal of Physiology: Regulatory, Integrative, and Comparative Physiology* 282 (2002): R1536–R1539.

CHAPTER 6: WHY NOT IGUANA?

166 *more than $10 billion worth of flavorings*: http://www.leffingwell.com/top_10.htm.

187 *Virgin Mary in a grilled-cheese sandwich*: Jessica Firger, "See the Virgin Mary on Toast? No, You're Not Crazy," *CBS News*, May 4, 2014, http://www.cbsnews.com/news/see-the-virgin-mary-on-toast-no-youre-not-crazy/. The phenomenon has been studied scientifically: Jiangang Liu et al., "Seeing Jesus in Toast: Neural and Behavioral Correlates of Face Pareidolia," *Cortex* 53 (2014): 60–77.

191 *many other languages have fewer*: Joseph Henrich, *The Secret of Our Success: How Culture Is Driving Human Evolution, Domesticating Our Species, and Making Us Smarter* (Princeton, NJ: Princeton University Press, 2016), 240.

194 *Australian chemistry teacher*: See, for example, James Kennedy, "Ingredients of an All-Natural Banana," https://jameskennedymonash.wordpress.com/2013/12/12/ingredients-of-an-all-natural-banana/.

196 *"the Dorito effect"*: Mark Schatzker, *The Dorito Effect: The Surprising New Truth about Food and Flavor* (New York: Simon & Schuster, 2015).

198 *"note-by-note cooking"*: Hervé This, *Note-by-Note Cooking: The Future of Food* (New York: Columbia University Press, 2014).

198 *BBC news report*: "Is This What We'll Eat in the Future?", video of Hervé This, BBC News, November 6, 2013, http://www.bbc.com/news/magazine-24825582.

198 *told one reporter*: Wendell Steavenson, "Hervé This: The World's Weirdest Chef," *Prospect*, September 2014, http://www.prospectmagazine.co.uk/features/herve-this-the-worlds-weirdest-chef.

199 *"dirac"*: Hervé This, "Three Recipes for Note by Note Cooking," *La Cuisine Note à Note* (blog), November 20, 2014, http://hthisnoteanote.blogspot.ca/2014/11/three-recipes-for-note-by-note-cooking.html.

CHAPTER 7: THE KILLER TOMATO

205 *40 percent lower than they used to be*: Donald R. Davis, "Declining Fruit and Vegetable Nutrient Composition: What Is the Evidence?" *HortScience* 44 (2009): 15–19.

206 *chose sixty-six varieties*: Denise Tieman et al., "The Chemical Interactions Underlying Tomato Flavor Preferences," *Current Biology* 22 (2012): 1035–1039.

206 *twice as sweet*: Linda M. Bartoshuk and Harry J. Klee, "Better Fruits and Vegetables through Sensory Analysis," *Current Biology* 23 (2013): R374–R378.

208 *to essential human nutrients*: Stephen A. Goff and Harry J. Klee, "Plant

Volatile Compounds: Sensory Cues for Health and Nutritional Value?" *Science* 311 (2006): 815–819.

210 *tested what made for a tasty tomato*: Tieman et al., "Tomato Flavor Preferences."

213 *looked at the volatiles in strawberries*: Michael L. Schwieterman et al., "Strawberry Flavor: Diverse Chemical Compositions, a Seasonal Influence, and Effects on Sensory Perception," *PLoS One* 9 (2014): e88446, doi:10.1371/journal.pone.0088446.

213 *a single gene variant*: Alan H. Chambers et al., "Identification of a Strawberry Flavor Gene Candidate Using an Integrated Genetic-Genomic-Analytical Chemistry Approach," *BMC Genomics* 15 (2014): 217, doi:10.1186/1471-2164-15-217.

217 *description that emerges*: Wendy V. Parr et al., "Perceived Minerality in Sauvignon Wines: Influence of Culture and Perception Mode," *Food Quality and Preference* 41 (2015): 121–132.

218 *ten times as much thiol*: W. V. Parr et al., "Association of Selected Viniviticultural Factors with Sensory and Chemical Characteristics of New Zealand Sauvignon Blanc Wines," *Food Research International* 53 (2013): 464–475.

218 *trucking the grapes*: Dimitra L. Capone and David W. Jeffery, "Effects of Transporting and Processing Sauvignon Blanc Grapes on 3-Mercaptohexan-1-ol Precursor Concentrations," *Journal of Agricultural and Food Chemistry* 59 (2011): 4659–4667.

218 *unique microbial ecosystem*: Nicholas A. Bokulich et al., "Microbial Biogeography of Wine Grapes Is Conditioned by Cultivar, Vintage, and Climate," *Proceedings of the National Academy of Sciences* 111 (2014): E139–E148, doi:10.1073/pnas.1317377110.

219 *detectably different aroma profile*: Sarah Knight et al., "Regional Microbial Signatures Positively Correlate with Differential Wine Phenotypes: Evidence for a Microbial Aspect to Terroir," *Scientific Reports* 5 (2015): 14233, doi:10.1038/srep14233.

221 *some store better than others*: Luke Bell et al., "Use of TD-GC-TOF-MS to Assess Volatile Composition during Post-Harvest Storage in Seven Accessions of Rocket Salad (Eruca sativa)," *Food Chemistry* 194 (2016): 626–636.

222 *vanished after a week*: Fernando Vallejo, Francisco Tomás-Barberán, and Cristina García-Viguera, "Health-Promoting Compounds in Broccoli as Influenced by Refrigerated Transport and Retail Sale Period," *Journal of Agricultural and Food Chemistry* 51 (2003): 3029–3034.

222 *haven't reached consensus*: See, for example, Marcin Baranski et al., "Higher Antioxidant and Lower Cadmium Concentrations and Lower Incidence of Pesticide Residues in Organically Grown Crops: A Systematic Literature Review and Meta-Analyses," *British Journal of Nutrition* 112 (2014): 794–811; Diane Bourn and John Prescott, "A Comparison of the Nutritional Value, Sensory Qualities, and Food Safety of Organically and Conventionally Produced Foods," *Critical Reviews in Food Science and Nutrition* 42 (2002): 1–34; Alan D. Dangour et al., "Nutritional Quality of Organic Foods: A Systematic Review," *American Journal of Clinical Nutrition* 90 (2009): 680–685; Crystal Smith-Spangler et al., "Are Organic Foods Safer or Healthier Than Conventional Alternatives? A Systematic Review," *Annals of Internal Medicine* 157 (2012): 348–366.

223 *local might not mean fresher*: I owe this idea to Alyson Mitchell of the University of California, Davis.

223 *didn't matter one bit*: Xin Zhao et al., "Consumer Sensory Analysis of Organically and Conventionally Grown Vegetables," *Journal of Food Science* 72 (2007): S87–S91.

224 *thought the eco-friendly coffee*: Patrik Sörqvist et al., "Who Needs Cream and Sugar When There Is Eco-Labeling? Taste and Willingness to Pay for 'Eco-Friendly' Coffee," *PLoS One* 8 (2013): e80719, doi:10.1371/journal.pone.0080719.

226 *says one tomato grower*: Quoted in Dan Charles, "How the Taste of Tomatoes Went Bad (and Kept on Going)," *NPR All Things Considered*, June 28, 2012, accessed March 1, 2016, http://www.npr.org/sections/thesalt/2012/06/28/155917345/how-the-taste-of-tomatoes-went-bad-and-kept-on-going.

227 *20 percent less sugar*: Ann L. T. Powell et al., "Uniform ripening Encodes a Golden 2-like Transcription Factor Regulating Tomato Fruit Chloroplast Development," *Science* 336 (2012): 1711–1715.

CHAPTER 8: THE CAULIFLOWER BLOODY MARY AND OTHER CHEFLY INSPIRATIONS

233 *more of the flavor in the vegetable*: Royal Society of Chemistry, "Kitchen Chemistry: The Chemistry of Flavour," http://www.rsc.org/learn-chemistry/resource/res00000816/the-chemistry-of-flavour.

234 *Researchers in England*: D. S. Mottram and R. A. Edwards, "The Role of Triglycerides and Phospholipids in the Aroma of Cooked Beef," *Journal of the Science of Food and Agriculture* 34 (1983): 517–522.

235 *skatole*: Peter K. Watkins et al., "Sheepmeat Flavor and the Effect of Different Feeding Systems: A Review," *Journal of Agricultural and Food Chemistry* 61 (2013): 3561–3579.

237 *621 different Maillard products*: Donald S. Mottram and J. Stephen Elmore, "Control of the Maillard Reaction during the Cooking of Food," in Donald S. Mottram and Andrew J. Taylor, eds., *Controlling Maillard Pathways to Generate Flavors* (Washington, DC: American Chemical Society Symposium Series 1042, 2010), 143–155.

239 *lab version of a cook-off*: Chris Kerth, "Determination of Volatile Aroma Compounds in Beef Using Differences in Steak Thickness and Cook Surface Temperature," *Meat Science* 117 (2016): 27–35.

240 *different sets of microbes*: My discussion of cheese styles follows Julie E. Button and Rachel J. Dutton, "Cheese Microbes," *Current Biology* 22 (2012): R587–R589.

244 *more popular than you'd expect*: Michael A. Nestrud, John M. Ennis, and Harry T. Lawless, "A Group-Level Validation of the Supercombinatoriality Property: Finding High-Quality Ingredient Combinations Using Pairwise Information," *Food Quality and Preference* 25 (2012): 23–28.

247 *experimenting with salty ingredients*: Heston Blumenthal, "Weird but Wonderful," *The Guardian*, May 4, 2002.

248 *Chartier's book*: François Chartier, *Taste Buds and Molecules* (Hoboken, NJ: Wiley, 2012).

250 *studied their molecular overlaps*: Yong-Yeol Ahn et al., "Flavor Network and the Principles of Food Pairing," *Scientific Reports* 1 (2011): 196, doi:10.1038/srep00196.

251 *The Flavor Bible*: Karen Page and Andrew Dornenburg, *The Flavor Bible: The Essential Guide to Culinary Creativity, Based on the Wisdom of America's Most Imaginative Chefs* (New York: Little, Brown, 2008).

257 *a Wassily Kandinsky painting*: Charles Michel et al., "A Taste of Kandinsky: Assessing the Influence of the Artistic Visual Presentation of Food on the Dining Experience," *Flavour* 3 (2014): 7, doi:10.1186/2044-7248-3-7.

259 *"a quick run through the blender"*: Nathan Myhrvold with Chris Young and Maxime Bilet, *Modernist Cuisine: The Art and Science of Cooking, Volume 4* (Bellevue, WA: The Cooking Lab, 2011), 343.

EPILOGUE: THE FUTURE OF FLAVOR

261 *craft breweries increasing by nearly 20 percent*: Brewer's Association, "Number of Breweries and Brewpubs in U.S.," accessed May 28, 2016, https://www.brewersassociation.org/statistics/number-of-breweries/.

261 *more than doubling since 2000*: Rebecca Smithers, "Good Beer Guide 2015 Shows UK Has Most Breweries per Head of Population," *The Guardian*, September 11, 2014, http://www.theguardian.com/lifeandstyle/2014/sep/11/good-beer-guide-2015-uk-most-breweries-per-head-population.

262 *Crisco white sauce*: Leslie Brenner, *American Appetite: The Coming of Age of a Cuisine* (New York: Avon, 1999), 21.

264 *The number of farmers' markets*: Anonymous, "Number of US Farmers' Markets Continues to Rise," accessed May 28, 2016, http://www.ers.usda.gov/data-products/chart-gallery/detail.aspx?chartId=48561&ref=collection&embed=True.

268 *found no difference*: Wendy V. Parr, David Heatherbell, and K. Geoffrey White, "Demystifying Wine Expertise: Olfactory Threshold, Perceptual Skill and Semantic Memory in Expert and Novice Wine Judges," *Chemical Senses* 27 (2002): 747–755.

268 *slightly more likely*: Gary J. Pickering, Arun K. Jain, and Ram Bezawada, "Super-Tasting Gastronomes? Taste Phenotype Characterization of Foodies and Wine Experts," *Food Quality and Preference* 28 (2013): 85–91.

INDEX